Advances in Industrial Control

Springer
London
Berlin
Heidelberg
New York
Barcelona
Budapest
Hong Kong
Milan
Paris
Santa Clara
Singapore
Tokyo

Other titles published in this Series:

Intelligent Seam Tracking for Robotic Welding
Nitin Nayak and Asok Ray

Identification of Multivariable Industrial Process for Simulation, Diagnosis and Control
Yucai Zhu and Ton Backx

Nonlinear Process Control: Applications of Generic Model Control
Edited by Peter L. Lee

Microcomputer-Based Adaptive Control Applied to Thyristor-Driven D-C Motors
Ulrich Keuchel and Richard M. Stephan

Expert Aided Control System Design
Colin Tebbutt

Modeling and Advanced Control for Process Industries, Applications to Paper Making Processes
Ming Rao, Qijun Xia and Yiquan Ying

Modelling and Simulation of Power Generation Plants
A.W. Ordys, A.W. Pike, M.A. Johnson, R.M. Katebi and M.J. Grimble

Model Predictive Control in the Process Industry
E.F. Camacho and C. Bordons

H_∞ Aerospace Control Design: A VSTOL Flight Application
R.A. Hyde

Neural Network Engineering in Dynamic Control Systems
Edited by Kenneth Hunt, George Irwin and Kevin Warwick

Neuro-Control and its Applications
Sigeru Omatu, Marzuki Khalid and Rubiyah Yusof

Energy Efficient Train Control
P.G. Howlett and P.J. Pudney

Hierarchical Power Systems Control: Its Value in a Changing Industry
Marija D. Ilic and Shell Liu

System Identification and Robust Control
Steen Tøffner-Clausen

Genetic Algorithms for Control and Signal Processing
K.F. Man, K.S. Tang, S. Kwong and W.A. Halang

Advanced Control of Solar Plants
E.F. Camacho, M. Berenguel and F.R. Rubio

Control of Modern Integrated Power Systems
E. Mariani and S.S. Murthy

E. Mariani and S.S. Murthy

Advanced Load Dispatch for Power Systems

Principles, Practices and Economies

With 72 Figures

Springer

E. Mariani
ENEL Div Trasmissione, via P.E. Imbriani 42, 80132 Napoli, Italy

S.S. Murthy
Consultant (Power & Energy), 145 7th Main 5th Block,
Jayanagar, Bangalore 560041, India

British Library Cataloguing in Publication Data
Mariani, Ezio
Advanced load dispatch for power systems : principles, practices and economies. - (Advances in industrial control)
1.Electric power systems - Load dispatching
I.Title II.Murthy, Surabhi Srinivasa
621.3'19

Library of Congress Cataloging-in-Publication Data
Mariani, E. (Ezio), 1932-
Advanced load dispatch for power systems : principles, practices, and economies / E. Mariani and S.S. Murthy.
p. cm. - - (Advances in industrial control)
Includes bibliographical references and index.

1. Interconnected electric utility systems- -Automation.
2. Electric power systems- -Load dispatching. 3. Electric power systems- -Control. I. Srinivasa Murthy, S. II. Title. III. Series.
TK447.M37 1997 97-15159
333.793'2- -dc21 CIP
ISBN-13: 978-1-4471-1251-8 e-ISBN-13: 978-1-4471-0991-4
DOI: 10.1007/978-1-4471-0991-4

Typesetting: Camera ready by authors

69/3830-543210 Printed on acid-free paper

Advances in Industrial Control

Professor Dr -Ing M. Thoma
Westermannweg 7
D-30419 Hannover
Germany

Professor H. Kimura
Department of Mathematical Engineering and Information Physics
Faculty of Engineering
The University of Tokyo
7-3-1 Hongo
Bunkyo Ku
Tokyo 113
Japan

Professor A.J. Laub
Department of Electrical and Computer Engineering
University of California
Santa Barbara
California 93106
United States of America

Professor J.B. Moore
Department of Systems Engineering
The Australian National University
Research School of Physical Sciences
GPO Box 4
Canberra
ACT 2601
Australia

Dr M.K. Masten
Texas Instruments
2309 Northcrest
Plano
TX 75075
United States of America

Professor Ton Backx
AspenTech Europe B.V.
De Waal 32
NL-5684 PH Best
The Netherlands

SERIES EDITORS' FOREWORD

The series *Advances in Industrial Control* aims to report and encourage technology transfer in control engineering. The rapid development of control technology impacts all areas of the control discipline. New theory, new controllers, actuators, sensors, new industrial processes, computer methods, new applications, new philosophies..., new challenges. Much of this development work resides in industrial reports, feasibility study papers and the reports of advanced collaborative projects. The series offers an opportunity for researchers to present an extended exposition of such new work in all aspects of industrial control for wider and rapid dissemination.

In Europe, and soon in the United States, power system deregulation is becoming widespread. This involves the privatisation of former public power utilities and the creation of power markets. The United Kingdom has recently undergone this transformation and the countries of the European Union are being encouraged to follow this deregulation policy. This volume *Advanced Load Dispatch for Power Systems* and its companion volume *Control of Modern Integrated Power Systems* both by Professor E. Mariani and Professor S.S. Murthy are therefore very timely additions to the power system literature and to the *Advances in Industrial Control* series.

Load dispatch is an operational issue, and this volume covers the technology and the economic and human factors involved in the problem. The volume gives an in-depth view of important qualitative concepts like security and reliability of the power system and describes the economic dimensions and the related technology for modern power system operations. Useful case study material based on different international experience is also described. The sum total is a volume likely to be of great interest to power system and control engineers alike.

M.J. Grimble and M.A. Johnson
Industrial Control Centre
Glasgow, Scotland, UK

FOREWORD

The book on Advanced Load Dispatch for Power Systems, written by Prof. E. Mariani and Prof. S.S. Murthy, is the first of its kind on the subject and deserves to be widely welcomed.

Prof. Mariani through his long association with ENEL brings with him the experience of a power system operating as part of the well-developed UCPTE in Europe while Prof. Murthy has been associated with the development of regional grid systems in India right from the inception and therefore brings with him a unique experience in developing interconnected operation of adjoining systems each with its own characteristics and methods of operation.

True, the well-developed systems in Europe and North America have made considerable advances in the art and science of operating the systems together satisfactorily with well-defined criteria, and thanks to the rapid development of reliable communication facilities and computer systems for efficient operation and control, economies of scale are being realised and reliability levels are continually increasing. On the other hand, in the developing systems, while the various concepts relating to system operation and control are well understood, in view of the load demands growing much faster than additions to generating capacity and the associated networks, it has not been possible to maintain satisfactory system security levels.

The authors have dealt with various aspects of load dispatch in a systematic and comprehensive manner. The balancing of available generation – hydro, thermal and nuclear – with the system demands which vary with the day and the season in the most economical manner, is a delicate task which has to be performed with due regard to economy as well as security of the system. The methodologies followed in such operational planning, that spread over a period of one to five years ahead, are well covered in the book. The various related aspects like telecommunication facilities required, operating reserves, load management and inter-system exchanges, are dealt with in detail.

The book would serve the needs of system planning and operating engineers , advanced students and researchers, and enable them to get a total view of the various aspects involved in system operation.

I am glad of the opportunity the authors have given me to go through their manuscript and write a Foreword. I have made a few suggestions and comments here and there which the authors have accepted and incorporated in the text.

L. PARIS

PREFACE

The first of us has been associated with a developed system which is part of the UCPTE system in Western Europe, interconnected with the systems in Eastern Europe through D.C. links (now also through A.C.).

The second of us, through association with developing systems, has seen how they have grown - first as individual power systems, next with interconnections developed between systems at frontier points, then as regional systems with 4-5 constituents and later on with links, A.C. or back-to-back D.C., established between regional systems, thus paving the way for a national unified power system. Regular load dispatch centres were functioning in two or three utilities only in the beginning. Later on, with the concept of regional grids gaining ground, interim regional load dispatch centres with some minimum facilities were established for co-ordinating with State dispatch centres which also were coming up side by side. Then followed the plans for permanent load dispatch-control centres with modern telemetering, computer facilities, etc.

We have had occasions to study in depth the manner in which the power systems and control centres have developed in Europe, North America, the former Soviet Union and in other parts of the world.

No systems are alike nor are the control centres and operating philosophies alike. They have their own individual characteristics and 'personalities'. Nevertheless, a common approach is evolving in system planning and operation thanks to the modern concepts of system analysis, computerisation, etc., through which the possibilities of realising economies of scale and reliability of operation are emerging.

We have attempted, through our joint endeavour, to cover the various aspects of load dispatch and system control, while keeping in mind the need for developing systems to evolve the operating philosophies in a systematic manner with reliability and economy as the twin goals. We have tried to cover the various technical, economic and commercial aspects in the operation of interconnected power systems and bridge the gap as between the developing systems and the well-developed power systems. If this book could help the advanced students of power system engineering, operating engineers at control centres, and the management of utilities, in some manner towards the realisation of the above goals, we would consider our efforts rewarded.

We are indebted to Prof. Luigi Paris, Professor of Power System Analysis at the University of Pisa (Italy), an eminent authority in power system planning, who pioneered the international 1000 kV R&D Project in Italy, for writing the foreword to the book. Our sincere thanks are also due to him for going through the manuscript patiently and offering useful comments and suggestions which we have taken into account while finalising the material.

Ezio Mariani **S. S. Murthy**

CONTENTS

CHAPTER 1

INTRODUCTION

1.1 Development of Interconnected Power Systems

In the early stages of electric power development, the supply systems were small and their operation was simple. As the demand for electrical energy increased, the power systems also expanded. Many of the large systems in turn were interconnected in order to derive the various technical and economic advantages inherent in the co-ordinated operation of power systems. It is now the established practice in every country to unify the power systems and operate them to mutual benefit.

Interconnections with neighbouring countries have also come into being. Thus, we have the UCPTE (Union for the Co-ordination of Production and Transmission of Electrical Energy) covering twelve countries in western Europe with an overall peak load of about 250 GW and energy consumption of 1500 TWh (year 1991), with a D.C. submarine cable link with the United Kingdom and D.C. back-to-back links with Eastern Europe. The UCPTE interconnection is now expanding eastward, with AC links with CENTREL (Poland, Ceka, Slovakia and Hungary) and with projects of reaching Romania and Bulgaria, at the moment of writing. Similarly, in North America (USA and Canada), there are four synchronous interconnections (the biggest one comprising about 100 individual systems), interconnected amongst themselves through D.C. links. The overall peak load of the North American system was about 650 GW and the energy consumption 3500 TWh in the year 1991.

1.2 Benefits of Operation of Interconnected Power Systems

The various benefits of interconnected operation of contiguous power systems may be summarised thus:

1. Reduction in generating capacity due to the diversity of load demands
2. Reduction in standby capacity
3. Increase in the size of generating sets, thereby reducing the capital and operating costs
4. Optimum utilisation of the available plant capacity and transmission facilities
5. Reliability of power supply
6. Improvement in frequency

1.2.1 Reduction in generating capacity due to the diversity of load demands

The daily peak demands of any two contiguous systems do not generally occur during the same hour of the day. As a result of this diversity, the maximum load of the combined system on a day would be less than the sum of the maximum demands of the individual systems on the same day.

Further, the maximum demands of different systems may occur not only at different hours of the day but also on different days of the month. This gives rise to the monthly diversity which would normally be higher than the daily diversity during the same month. In addition to the daily and monthly diversities, there is also the annual diversity, which is given by the difference between the sum of annual peak demands of the constituent systems and the annual peak demand of the combined system.

The daily diversity brings about an improvement in the daily load factor of the composite system, which is therefore of immediate operational advantage and effects savings in the operating and installed capacities. A classic example is the interconnection between the power systems of Great Britain and France. Peak load hours in these countries are far enough apart, so that they can assist each other and thus save large additional outlays involved in meeting the peaks of the respective systems.

As a result of the monthly diversity, the monthly peak demands of the interconnected system would be less than the sum of the individual system peaks, thereby increasing the capacity available for maintenance and reserve in the month.

The annual diversity factor may be used to estimate the diversified annual maximum demand for a region, which would determine the generating capacity for the region. The annual diversity results in savings in installed capacity.

1.2.2 Reduction in standby capacity

In order to ensure dependable supply of power to the consumers, reserves have to be maintained over and above the capacity required to meet the anticipated peak demand. These reserves are required for carrying out the scheduled maintenance of generating plants and also to meet two major contingencies - unplanned outages occurring due to breakdown of generating units and uncertainties of load forecast. By sharing reserves through interconnections, a group of systems can reduce the combined reserve for forced outages since it is unlikely that maximum outages of units would occur at one time in all the systems.

Increasing use of larger and more economical units would further increase the importance of sharing of reserves as the individual systems may find it difficult to support heavy reserves necessitated by the use of large generating units.

1.2.3 Increase in the size of generating sets

With interconnected operation, larger generating stations with larger units, resulting in lower capital and operating costs, can be located at the most economical sites in the region as a whole. Individual systems would find it difficult to support such large stations particularly as base-load stations.

1.2.4 Optimum utilisation of the available plant capacity and transmission facilities

With interconnected operation, thermal stations can be used more effectively in conjunction with hydro-electric plants. During the high water periods, the hydro stations could be operated at higher load factors due to extra water available, thereby contributing more energy to the system, when some of the thermal stations could be shut down. Thus, on the one hand, it would be possible to utilise excess energy available from hydro stations which would otherwise go waste, and on the other, it would save large quantities of coal. In dry periods, the reverse would be the case. During this period, the thermal stations would be working at a higher load factor so that the water in reservoirs at hydro stations could be conserved. The interconnection of the hydro areas may also yield firm power because of possible diversity in water availability patterns. Thus it is possible to optimise the total generation available in the region.

1.2.5 Reliability of Power Supply

The contiguous systems can provide mutual assistance in case of emergency. Short-term emergencies occur in every system due to outages of generating plant, substation equipment and transmission facilities. There are numerous instances of emergency assistance being provided by contiguous systems. A classic example is that of closure of five out of eight generating units at the Sharavathi Power Station in the State of Karnataka, India, for fifteen days in August 1972, necessitated by tunnel repairs which were to be carried out urgently. This meant a loss of generation of 450 MW in that system. The State of Andhra Pradesh also had a shortage at that time. However, rescheduling was done in the Andhra Pradesh, Tamil Nadu, Kerala and even Maharashtra systems in such a manner that the major portion of the deficit was met, thanks to the co-operation of these four systems.

1.2.6 Improvement in Frequency

In a system operating independently, the frequency fluctuates continuously in small amplitudes whereas when the same system is interconnected to a large grid system, the combined system frequency is more steady. The reason for such frequency behaviour is due to the fact that when a system is operating independently, even a small change in its load is sensed by the system, whereas when this system is operating interconnected with one or more systems, the small variations of load in its own system become small or negligible as compared to the total load of all the systems interconnected.

1.3 Developing and Developed systems

Two examples are cited here, one relating to a developing country (India) and the other to a group of developed countries interconnected amongst themselves in Europe (UCPTE).

1.3.1 India

In India, the concept of interconnecting the power systems and operating them on an integrated basis was recognised in the 1970s. Accordingly, for purposes of planning and operation, the country was divided into the following five regions paving the way for an All-India Grid System (Table 1).

Table 1: *Operational Regions - India*

Regions	*Constituent States*
Northern	Jammu & Kashmir, Himachal Pradesh, Punjab, Haryana, Rajasthan, Delhi and Uttar Pradesh
Western	Gujarat, Madhya Pradesh, Maharashtra, Goa, Daman & Diu
Southern	Andhra Pradesh, Tamil Nadu, Karnataka, Kerala and Pondicherry
Eastern	Bihar, West Bengal, Orissa, Sikkim and Damodar Valley Corporation
North-Eastern	Assam, Meghalaya, Arunachal Pradesh, Nagaland, Manipur, Tripura and Mizoram

The above demarcation of the country into five regions has the regional boundaries conforming to the geographical boundaries of the States. While this might be considered as a practical arrangement to start with, now that a number of years have elapsed and operational experience is available , a review of the present demarcations would be desirable from the points of view of economy and operational efficiency. An optimum demarcation should take into account the following:

I. The best use of diversity,
II. Load-generation balance on a long-term basis,
III. Economic utilisation of transmission facilities, and
IV. Operational convenience.

In this process, part of one State may fall into one geographical region and part into another region. However, this need not be considered as a drawback. Economy and operational efficiency should be the guiding factors.

In India, power stations and transmission networks are built and operated not only by the States within the respective boundaries but also by the Central Government undertakings in the various regions for the benefit of the constituent States. Thus, hydro-electric stations are built and operated by the National Hydro-electric Power Corporation, thermal stations by the National Thermal Power Corporation and atomic power stations by the Nuclear Power Corporation of India. The power from these stations is shared in a certain proportion by the constituent States in each Region.

The role of the Central Sector Undertakings is increasing from year to year as will be evident from Table 2.

Table 2: *Relative role of power supply utilities - India*

	Installed Capacity (MW)	
	1988-89	**1994-95**
State Electricity Boards	42,220	52,854
Central Sector Undertakings	13,637 (23.5)	24,766 (30.5)
Private Sector Utilities	2,155	3,544
Total	**58,012**	**81,164**

(Figures in brackets indicate percentage of total).

In 1989, the Power Grid Corporation of India Limited (PGCIL) was formed with the objectives of strengthening regional power grids and establishing inter-regional links, ensuring efficient operation and maintenance of transmission systems under its charge, and establishing or augmenting Regional and National Load Dispatch Centres and communication facilities.

1.3.2 UCPTE

The UCPTE was founded in 1951 with representatives drawn from the electricity supply utilities of Belgium, Western Germany, France, Italy, Luxembourg, Holland, Austria and Switzerland. The UCPTE now includes not only these eight founder-member countries but also four Associate member-countries - Spain, Portugal, Yugoslavia and Greece and the four countries of CENTREL.

The objective of the UCPTE has been to contribute to the best possible extent the utilisation of all installations, existing or planned, for production and transmission of electric power in the member-countries and towards this end, to facilitate and expand international energy exchange. Thus, while in 1947, about 2 percent of the total energy consumed was transmitted over the links interconnecting some countries of Western Europe (even when the systems were not fully integrated within those countries), in 1992, about 7 percent of the total energy consumed in the UCPTE countries was exchanged amongst them.

Apart from the promotion of power exchanges, the UCPTE has also made efforts to reduce spill energy with continued success. Thus, the spill energy was reduced from 5 per cent in 1957 to 0.1 per cent in 1969.

It should be mentioned that, while frequency regulation, exchanges of power and energy, and co-ordination of communications facilities and associated problems come within the purview of the UCPTE, it is not concerned with the contracts that are executed between the member-countries for power exchange.

In the earlier years, power exchanges in UCPTE were handicapped by quota and currency restrictions in various countries. At present, however, prior authorisation is required only for long-term contracts, the administrative formalities in respect of occasional and seasonal supplies amongst UCPTE countries having been suspended once and for all. National frontiers have been fully abolished, so far as power transfers are concerned.

Let us have a brief look at the operation of these grid systems, i.e., the developing Indian grid systems and the well-developed UCPTE grid system.

1.3.3 Indian Grid Systems

A summary of press reports of happenings in the Southern Regional Grid in April and May 1994 and in April 1995 is given below which brings out the system operational deficiencies.

"The Hindu", April 10, 1994

> *One State or another in the regional grid often overdraws power in excess of its allocated quota (which varies from day to day depending on the availability in the Central pool) to meet excess demand not satisfied by the concerned State Electricity Board's own generation and the day's Central quota. Such excessive withdrawals, especially during peak periods, lead to the frequency in the system dropping abnormally below 50 cycles per second, sometimes down to 48 or 47, posing a threat to sensitive and costly power generation and transmission equipment in the entire region and upsetting electronic controls and communications. This triggers a chain of tripping of plants.*
>
> *To neutralise part of the overdrawal on their account, the erring States 'dump' power generated in their own power plants into the grid during low-load hours, normally between 10 p.m. and 6 a.m., boosting frequency well over the normal 50 cycles, again endangering power equipment and transmission systems and triggering grid disturbance and trippings.*

"The Hindu", May 11, 1994

> *The frequent load shedding, both scheduled and unscheduled, by the Karnataka Electricity Board in the last two months has been attributed to the poor power management by Andhra Pradesh State Electricity Board (APSEB) which had resulted in heavy drawing of power by Andhra Pradesh from the Central power generating stations.*
>
> *According to official sources, this trend of overdrawal of power had continued since three years, in spite of repeated requests by the affected member states of the grid. It was pointed out that Andhra Pradesh could have managed the power shortage, if it had resorted to load shedding and other measures to conserve power. But it had not done so.*
>
> *The APSEB was also asked to step up its own generation and maintain the frequency, which was coming down. But, on the contrary, the APSEB had continued to overdraw.*
>
> *The result of this was low frequency and low voltage in Karnataka and Kerala and even total black-out for a day in Tamil Nadu.*

"The Hindu", April 9, 1995

> *There were eight grid collapses between March 16 and March 31, because of overdrawal of power by Andhra Pradesh during the low voltage and low frequency regime. Both Karnataka and Tamil Nadu were suffering for no fault of theirs and badly affected by the blackouts during the breakdown, official sources said.*

Similar happenings were reported in other regions also from time to time.

1.3.4 The UCPTE system

The UCPTE system spans sixteen countries working satisfactorily with well-defined criteria which are scrupulously adhered to by each member-country. In broad terms, these criteria are:

a) A well-forecasted daily load curve for each system with a generation programme (plus imports from or exports to the neighbouring systems) to match the load curve with appropriate spinning reserves (either available within the system or availed of from the neighbouring systems and paid for), and

b) A clear-cut strategy to meet possible emergencies; the overall system frequency is invariably held at 50 ± 0.05 Hz.

From the two examples cited above, the vast difference prevailing between developing power systems and well-developed systems is apparent; considerable effort is required to bridge the gap. A major objective of this book is to draw attention to the various aspects involved in building up a suitable infrastructure for a systematic resolution of the various problems involved in power system operation and control.

It often happens that certain deficiencies in the organisational set-up hinder effective operations. At present, the Power Grid Corporation of India Limited (PGCIL) owns and operates the transmission system interconnecting the Central power stations with the State grid systems and has a Regional Dispatch Centre in each Region, with plans to set up a National Dispatch Centre to co-ordinate and direct the regional and inter-regional operations at the national level.

However, from the reports cited above, it is apparent that the responsibilities of PGCIL in the day-to-day operation of the grid system require to be clearly defined and that a restructuring is called for.

The PGCIL could be modelled on the lines of the National Grid Company (NGC) in England and Wales. The NGC is in charge of operation and maintenance of the grid system and its development and interconnections with France and Scotland. By virtue of its responsibility of management of the pool, the NGC plays its role through giving its concurrence to the generation programme, and also controls the frequency and voltage of the grid system while meeting the system demand.

Like the NGC, the PGCIL could own and operate the entire trunk 400/220 kV grid system in each Region; and taking into account the respective shares of the constituent States in the Central and jointly-owned generating stations, draw up the Regional generation schedules in close association with all the concerned interests and obtain their concurrence for implementation, and operate the grid system with due regard to frequency and voltage.

1.4 Human Factors in the Operation of Interconnected Power Systems

The present century has witnessed a phenomenal growth in the electricity supply industry all over the world. The statistics of rates of increase in generation and sale of electricity in different countries in the past few decades are quite impressive. What is probably not as well known, nor well publicised, is the development of the science and art of power system control. Side by side with the growth in generation, transmission and distribution, system operation has also grown in stature almost beyond recognition. The isolated generating units, often at low and medium voltages, situated at the load centres in metropolitan cities, gave way to generating stations situated at a number of locations and high voltage transmission lines interconnecting these stations and the load centres. Then came the interconnections and power pool concept, which brought in its wake problems of fault MVA, stability, tie line control and many others.

It is now generally acknowledged that interconnected operation of contiguous power systems is aimed mainly at economic and effective utilisation of the available plant and resources. Some of the human qualities required for successful operation of an interconnected power system to achieve this objective are the following:

1.4.1 Mutual Trust

The interests of the interconnected system must come before the conflicting interests of individual constituents. Taking a short-term view should give way to joint working for the benefit of the entire region and the equitable sharing of such benefits.

1.4.2 Understanding problems peculiar to constituents

Just as different classes of loads are differently treated in any electricity undertaking, similarly, all the problems of different undertakings in the region cannot always be equated to the same formula. An adequate knowledge of the system as a whole with special emphasis on the peculiar problems faced by the constituents is essential for those responsible for interconnected system operation.

1.4.3 Effective communication

Telecommunication engineers can claim that progress in system control is linked up with the rapid strides in the field of communication engineering. But what has been applied to power system operation so far is only an insignificant portion of the recent developments on the subject of telecommunications. For any control system, however automatic and computerised it may be, there are bound to be human operators; effective functioning of the system will depend on effective communication between man and machine and between men. The need for adequate training of operating engineers in the art of effective communication cannot be overemphasised.

CHAPTER 2
OBJECTIVES, FUNCTIONS AND LOCATION OF LOAD DISPATCH CENTRES

2.1 Objectives

The responsibility of an electric supply utility is to ensure that, at all times,

- the quality of supply is as high as possible, that there is continuity of supply and the voltage and frequency of supply are within the permissible limits, and
- power is supplied to the consumers at the lowest cost possible.

Load Dispatch Centres are set up by the power utilities at appropriate locations with a view to assisting them in discharging these responsibilities.

The terms *Load Dispatch Centres* and *Control Centres* have been used rather synonymously in many places in the text. What used to be called Load Dispatch Centre in the past is now generally termed as Control Centre. The major distinction between the dispatching office or centre of the past and the control centre of recent years is the incorporation of functions related to system security. The addition of the Security Monitoring function to the traditional functions of generation dispatching and supervisory control and data acquisition (SCADA) results in system-wide data acquisition, computer system and colour CRT displays.

The modern Control Centre does perform both the dispatching and control functions as stated below:

- Dispatching
  - Operational Planning (long, medium and short-term scheduling) for power system operation including economic optimisation and security evaluation
  - Decision-making for the control actions on the power system, with reference to economy and security of the system itself
  - supervision of the system state
- Control
  - Supervision of the state of the elements (generators, transformers, line breakers, etc.) of the power system
  - Verification of the conditions for the execution of commands, and their actual execution

- Decision-making and execution concerning security of individual elements of the system and safety

2.2 Operational Planning

The operational planning of a power system covers normally three periods - the year, the week and the day, and covers the following aspects:

- Preparation of annual programmes taking into account
    - the management of hydraulic reservoirs
    - the maintenance of generating units and transmission networks
    - the provision of fuel, the minimum stocking of which is dependent on the security of supply required for the thermal stations taking exceptional circumstances like strikes, transportation bottlenecks, etc., into account
- Preparation of weekly programmes on a particular day, say, Tuesday, specifying:
    - the contribution of hydro (with due regard to the quantities of water which should be stocked or emptied out each day) and thermal generation
    - the exchanges with the neighbouring systems
    - the network structure
- Preparation of daily programmes, hour by hour or 15 min. by 15 min., for the next day, or determination of optimal generation schedules of thermal and hydro generating units in the 24-hour period of programming, covering
    - the anticipated load curve
    - the selection of generating units required to meet the hourly demand levels with sufficient spare capacity to meet emergency conditions, and issuing instructions to the stations so that the units are run as required
    - the planning of power transfers with the neighbouring systems
    - the network structure required to cope with the demand with the available generation
    - evaluation of the possible consequences taking likely contingencies into account and determining the best possible operating schemes in various circumstances.

The planning of operation of the system on an annual, weekly and daily basis is one aspect of dispatching. The second aspect relates to actual dispatching. In practice, there are bound to be variations in load anticipation, sudden outage of generating or transmission plant, etc. As and when these situations arise, steps have to be taken to rearrange the system suitably with necessary adjustments in generation or transmission facilities so that the demands are met and no circuit is overloaded by the emergency outage of plant. Further, the variations that arise between the anticipated load and the actual load give rise to deviations in the system frequency, and efforts have to be made to bring the frequency within the acceptable limits.

Thus, the objectives of load dispatching may be summarised as follows:

1. To ensure power supply to consumers at every instant, and at all points of the network, with the maximum security that the generation and transmission facilities can provide,
2. To limit the duration and extent of the repercussions due to faults, and restore normal functioning of the network with the utmost rapidity,
3. To secure the optimum utilisation of the available generation and transmission facilities with a view to minimising the operating costs, and
4. To regulate the frequency.

2.3 Functions

The functions of the Load Dispatching Centre may be grouped under the three main categories

I. Programming

II. System Monitoring

III. Control

2.3.1 Programming

2.3.1.1 Load Forecasting

The load forecasts and the scheduling of generating plant to meet the load are prepared on a weekly basis and corrected on a daily basis one day in advance. The forecasts should be made as accurately as possible with reference to the load curves of the previous day, the corresponding day in the previous week and year (after making suitable allowances for the load growth during the course of the year) and the latest weather forecast. Due allowance should be made for special events, if any, for that day. Forecasting errors should be held within a close range so that the distribution of load at the various generating stations can be done in the most economical manner. An example of good programming is provided in Figure 2-1, which compares the predicted and the actual daily loads of the ENEL system (Italy) for 18th of January, 1995.

A record of all variations from the anticipated load that occur in practice should be maintained and the various causes for the variations analysed by the Load Dispatching Office, with a view to improving the technique of forecasting in a systematic manner.

Daily load forecast is often performed heuristically by skilled human operators but in many cases they are supported by automatic algorithms, implemented on computer and based mainly on extrapolative methods, of which a large variety is available.

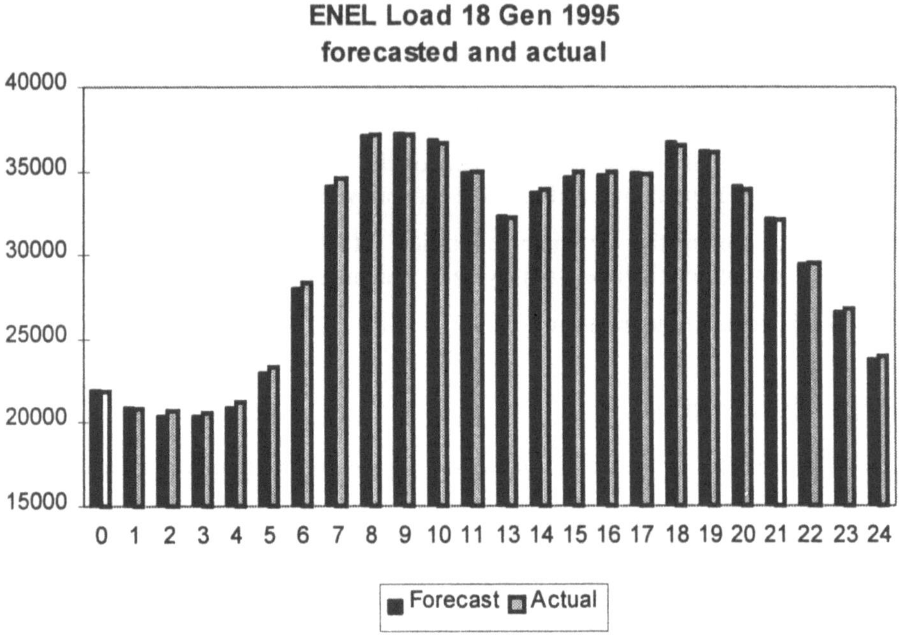

In recent years, experience has been acquired on implementation of such algorithms on artificial neural networks (ANN). The automatic algorithms, when available, work out historical series of load which undergo the particularities of the season and the weather, making due allowance for contingencies such as labour disputes, etc. This gives a base forecast, on which the operator may work taking into account the forecast contingencies, if any.

Appendix 1 gives the present practices on load forecasting based on a survey of 44 utilities carried out by CIGRE in 1991.

2.3.1.2 System reserve

A certain amount of reserve is always maintained in the power system in excess of the load demand to provide for overhaul and maintenance and also to provide running spare capacity. The latter is intended to provide regulating capacity, cover forced outages of generating plant and cover errors in estimating the load demands. The quantum of reserve depends on the load level, seasonal and daily variations in the load, maintenance requirements, number and size of generating units, forced outage rates of the sets, characteristics of the generating plants, limitations of plant capacity and the desired standard of reliability.

The subject of system reserve is dealt with in detail in Chapter 5 - *Determination of operating reserve.*

2.3.1.3 Daily generation scheduling

Plant scheduling for meeting the daily load demand is done in the manner shown in Fig. 2.2. The run-of-the-river plants fill up the bottom portion of the curve and the storage hydro plants fill up the top portion of the curve. The remaining portion of the curve is to be filled up (from bottom upwards) with the nuclear and high merit thermal plants and then with the relatively older thermal installations. The actual allocation of thermal generation shown in the curve to various power stations and units would of course need further sophistication on considerations of incremental generation costs, transmission line losses, etc. (See Section 2.1 of [1]- Economic load dispatch for thermal power plants).

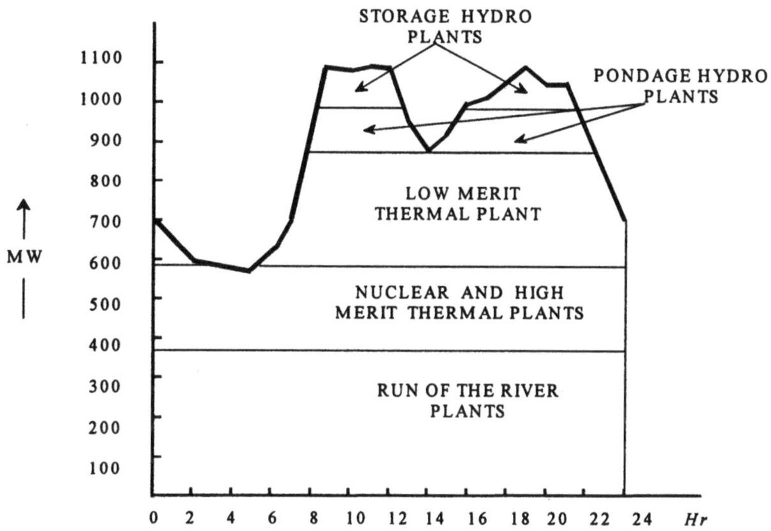

Figure 2.2: *Daily Generation Scheduling to cover a given load demand curve*

2.3.1.4 Reactive power

In power system planning as well as operation, kilovars are no less important than kilowatts. To maintain proper power flows, voltage levels and system stability and to minimise line losses, reactive generation is required at strategic locations. This requires a study of system reactive power characteristics and the nature and economics of various types of resources like generators, synchronous condensers and static capacitors.

Proper scheduling of reactive power for system operation requires provision of adequate on-line reactive power to meet the reactive load demand and certain reserve to cover forced or planned outage of a reactive generating source or a long transmission line or to meet the increase in reactive demand due to change in the weather conditions (e.g. air conditioning load entails lower power factors). The reserve is usually maintained at the generating stations where the generators can be over-excited up to the permissible field or stator current limits. Light-load periods pose some problems; receiving station voltages remain high even after lowering generating station voltages to the minimum allowable limit and after allowing for absorption of the maximum possible reactive power by generators, synchronous condensers, etc. Provision of reactors, disconnection of certain lines, etc., are some possible remedies (See also chapter 3 of [1]).

2.3.2 System Monitoring

In actual practice as the system conditions deviate from time to time with respect to the scheduled programme, the network has to be controlled in an appropriate manner. This is to ensure that the generation and load demands are satisfactorily matched and the active and reactive power flows, voltage levels, etc. at the various points of the network are satisfactory. The ability to take prompt and judicious decisions to cope with the situations arising out of any eventuality that may occur in the system becomes the 'touchstone' of the dispatcher.

In an interconnected network, the imbalances between the programmed generation and consumption do not generally exceed a small percentage of the programme established and are largely covered by the available spinning reserves. However, a sudden outage of power plant or an important feeding line, increase or decrease of load, etc., or a combination of these factors can lead to emergency situations.

It is therefore essential to have a predetermined plan to meet such emergencies. For this purpose continuous monitoring of the system is necessary. The control centre has therefore to know at any time:

1. the exact configuration of the transmission system
2. the active and reactive power flows in the lines and transformers and voltage conditions at important points in the system
3. the active and reactive power generation and load levels at different points
4. the interchange of power at the points of interconnection with the adjoining systems
5. the frequency at switching stations where systems are to be coupled.

All this information is picked up at different power stations and sub-stations and telemetered or telesignalled at the control centres, and updated every few seconds, thus giving an almost instantaneous picture of the network conditions.

Where a real time process control computer is installed, it receives all the data and checks their validity, e.g., by means of algorithms of state estimation which are run every few minutes. This information is displayed in five different ways:

- on the wall diagram which gives an overall view of the network;
- on strip chart recorders which display the magnitude and the trend of certain

important system parameters such as generation, tie-line loadings and frequency;

- on CRT displays which, through the use of a keyboard connected to the computer, display information such as the single line diagram of the stations, the load flow on the lines emanating from them, the voltage levels, the load-generation balance of the area, etc.
- on "alarm" CRTs displaying all the changes in the network topology (for example, outage of a line), or indicating the quantities that are out of preassigned upper and lower limits (for example, a line having reached 90% of its maximum rating);
- on high-speed printers or curve plotters which display information on request.

Thus, the control centre possesses all the data required for a real time assessment of the power system and taking such decisions as may be warranted.

2.3.3 System Control

When systems are tied together and are operating in parallel with one another, the entire system frequency is one. Any appreciable change of load or generation in a system will give rise to a change in the system frequency, and steps have to be taken to regulate the frequency so as to maintain it roundabout its nominal value.

When two or more systems are interconnected, in addition to the regulation of the common frequency, there will also be the question of regulation of power exchanges between the various systems. This regulation ensures the execution of the programmes of exchange of power in accordance with the commercial agreements. During normal operation, each system should discharge its regulating responsibilities and avoid unintentional and unscheduled transfers of power over interconnecting ties. Thus, the role of regulation is to compensate at every instant the inevitable difference between the actual load curve and the programmed load curve, as well as the difference between the actual and the scheduled generation. This regulation duty is entrusted to a certain number of power stations called 'regulating stations'. These regulating stations have a clear plus/minus range for variation of generation to take care of the changed load-generation balance. Each one of the regulating stations contributes a certain 'regulating band' representing the difference between the maximum output and the minimum output that is considered technically and economically suitable.

In comparatively small systems, it might suffice to assign the regulation of frequency to one power station. In larger interconnected systems, it becomes necessary to regulate the frequency and the flows on the tie lines, and assign the regulation duty to a number of stations. This is called load-frequency control. This term however is a misnomer in that one does not really control the load on the system, but rather controls the generation to exactly meet the load at the frequency of 50 or 60 Hz. and maintain the tie-line flows to scheduled values. However, the term 'load-frequency control' and the abbreviation LFC have come to stay. LFC is dealt with in detail in Section 2.2 of [1] - Load-frequency control.

Accordingly, the control actions under normal conditions and in emergency situations would be as under:

- On the operation of power stations

  If the load variations are small, the power output of 'controlled' stations automatically changes so as to constantly balance generation against load. When such automatic adjustments are not sufficient, the control centre issues instructions to modify the programme of certain power stations or to initiate the commitment of additional generating units or call for modifications of scheduled interchange programmes as may be possible.

- On the network configuration

  In normal operation, when overload limits are exceeded, either the network is restructured or generations are altered through instructions from the control centre to the sub-stations or power stations as the case may be.

- When emergency situations arise on account of failures in the power stations or on the transmission lines

  The role of the control centre then is to:

  - resort to load shedding in the extreme cases,
  - re-establish normal supplies to the customers affected,
  - eliminate overloads and restore the entire system to normal.

Emergency control is dealt with in some detail in Chapter 5 of [1].

The block diagram in Fig. 2.3 gives a layout of the various aspects to be considered in power system operation. Fig. 2.4 shows in a different but parallel way how some operational activities are linked with the behaviour of the power system and its control.

2.4 Hierarchical set-up of Load Dispatch Centres

In the initial stages when the system is relatively simple with a few power stations and transmission lines, it would generally suffice to have one Load Dispatch Centre at the most suitable location. This is the simplest case in which all the functions are performed in a single tier (Fig. 2.5a). As the system develops and becomes more and more complex, the dispatching-control system is structured in two or three tiers (Figs. 2.5 and 2.6).

The functions are distributed between the tiers, with the possibility, in many cases, of one tier serving as a back-up to another for a particular function, or co-operating in its implementation.

The functions may be grouped as follows:

A- data acquisition of the network state, data display,

B- data recording for setting up historical files;

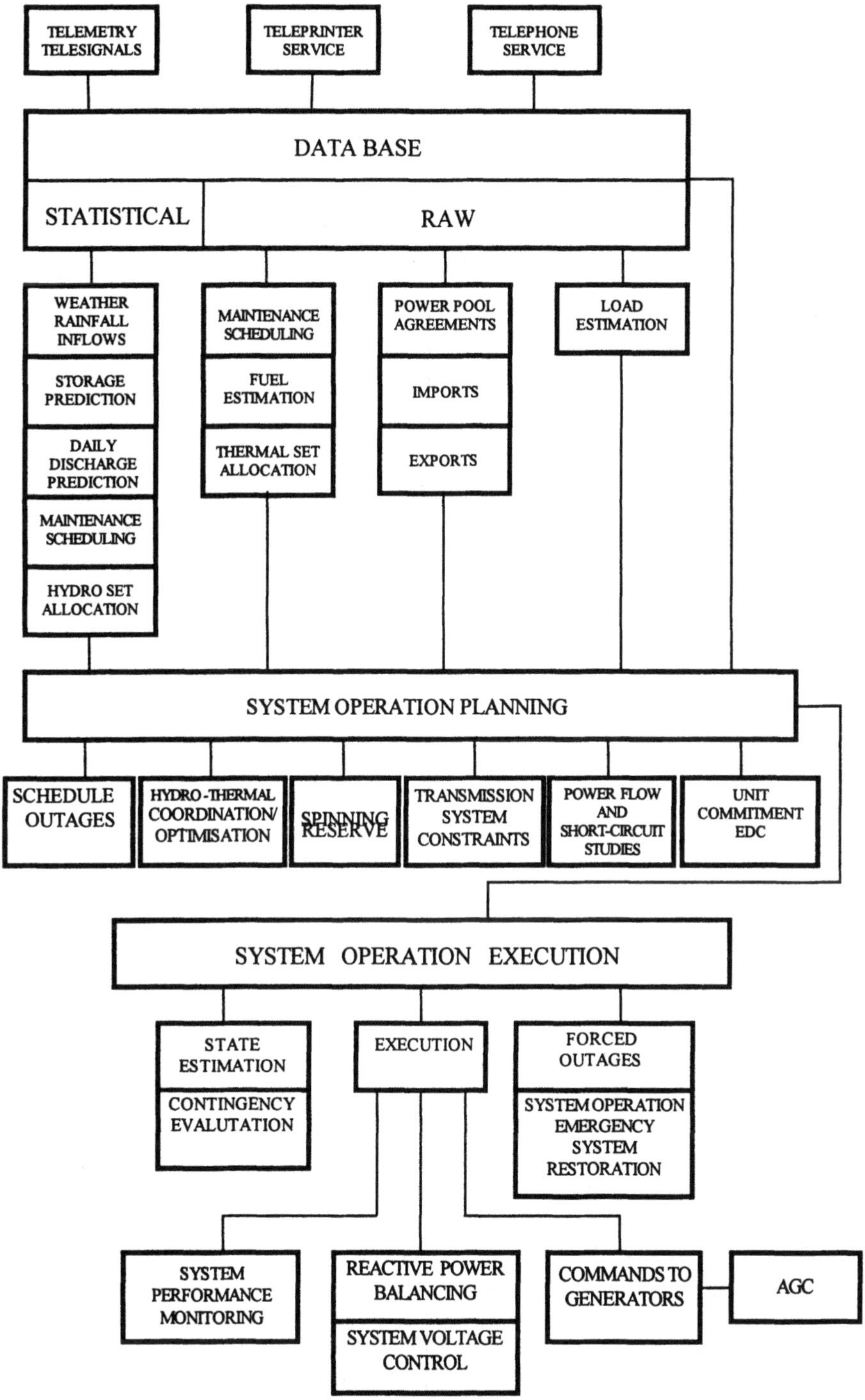

Figure 2.3: *Block Diagram of Power System Operation*

C- load-frequency control (LFC) or Automatic Generation Control (AGC);
D- emergency control (load shedding, start-up of peaking units, etc.);
E- remote control in normal conditions (breaker control);
F- other functions: state estimation, security monitoring, economic optimisation,
G- scheduling, off-line simulation.

Fig. 2.5 b) (two tiers) represents a typical case of one System Dispatching Centre performing A, B, C, E, F and G functions, with some Remote Control Centres, each supervising one part of the system and performing the functions A and D; the second tier Remote Control Centres come hierarchically under the first tier System Dispatching Centre.

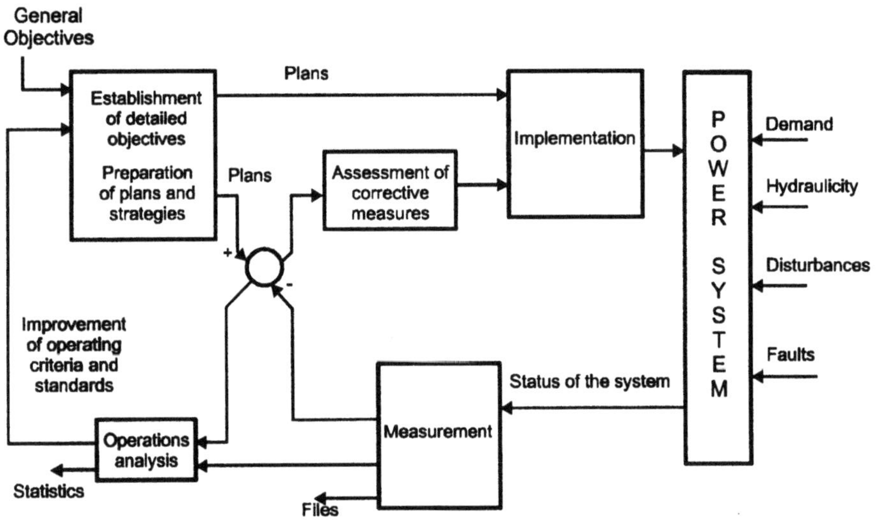

Figure 2.4: *General Control Scheme of an electrical power system to an independent set of inputs (faults, disturbance, etc.) and an action of scheduling and control.*

Fig. 2.5 c) (three tiers) is typical of large systems (e.g., more than 500 electrical nodes and 100 or more power stations covering a vast geographical territory), and consists of one System Dispatching Centre performing A, B, C, F and G functions, Regional or Area Control Centres performing A and B over a part of the system and coming hierarchically under the upper tier, and many Remote Control Centres, coming under Regional Control Centres, performing A, D, and E on a restricted part of the system.

As far as centripetal data flow (such as active and reactive powers, breaker status, etc., picked up from the plants and forwarded to control-dispatching centres) is concerned, one typical arrangement for the three -tiers scheme is the following: data are sent independently from plants to third tier (Remote Control) and second (Regional or Area Control) tier; the System Dispatching Centre (first tier) receives data from the Regional (second tier) centres. However, there is also the possibility of having the first tier fed directly with data from the field.

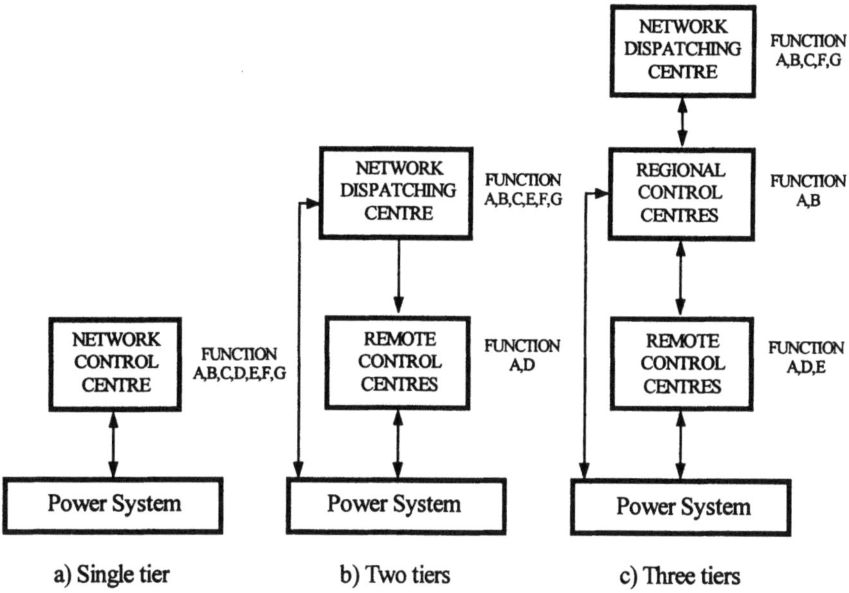

A - Acquisition of the network state and visualization
B - Data recording
C - LFC or AGC (load frequency control or automatic generation control)
D - Emergency control
E - Remote-control in normal conditions
F - Security monitoring ; economic optimi zation
G - Scheduling ; off-line simulation

Figure 2.5: *Structure of Dispatching-Control Systems*

Fig. 2.6 shows a simplified scheme of a three-tier structure, with on-line and off-line computer systems; off-line functions (like scheduling and post-mortem analysis) are performed only at the first and second tier levels.

Where there is more than one tier, there are the possibilities of partitioning the electrical system. One possibility is to entrust the control of the highest voltage level network (for example, 400 kV) for the whole territory to the first tier, and the next voltage network (for example, 220 and/or 132 kV) over a particular portion of the territory to the second tier.

More often, however, there is a redundancy: the first tier controls the whole network (for example, 400 and 220 kV) for the whole territory, and the second tier centres also cover the 400 and 220 kV networks of the respective territories; the second tier centres (regional) server as back-up to the first tier. In addition, the second tier may have the control of 132 kV network, which is generally ignored by the first tier.

In the event of failure in communications, channels or equipment at the main control centre, the Regional or Area Control Centres themselves will process the data acquired pertaining to their areas until the main control centre comes back into operation.

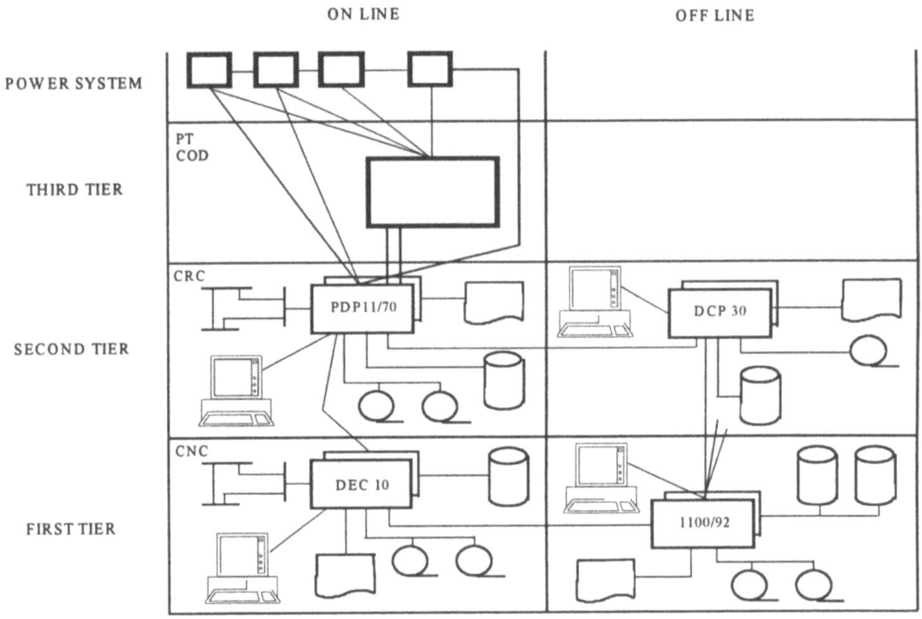

The following examples, two relating to modern systems (Great Britain and France) and one relating to a developing system (India) give the broad structural arrangement of Load Dispatch Centres.

2.4.1 Great Britain

There are nine Grid Control Centres operating and controlling the whole of the British Grid System. The North of Scotland Hydro-electric Board (NSHEB) and the South of Scotland Electricity Board (SSEB) have their own Grid Control Centres, and the Central Electricity Generating Board (CEGB) has seven Control Centres and a National Control situated in London which co-ordinates the working of the seven Control Centres. The power transfers between SSEB and NSHEB are on the basis of pre-arranged mutually agreed programmes, both parties trying to keep the transfers according to a schedule. Between SSEB and CEGB, the transfer of power is again on a mutually agreed programme. CEGB being the largest system is responsible for frequency regulation whereas SSEB regulates the load so as to keep the tie-line transfer appreciably constant.

Among the seven different regions of CEGB, the power transfer is frequency biased. The National Control directs the regional controls as to the magnitude and direction of the inter-regional transfer they must maintain.

The CEGB in England and Wales has since been privatised into two generation companies - Power Gen and National Power - in charge of thermal stations, one National Grid Company (NGC) and twelve distribution companies. Nuclear Electric, a public utility, will be in charge of nuclear power stations. Nevertheless the structure of the control remains centralised in the NGC.

2.4.2 France

For purposes of system operation, the country has been divided into eight Regions and a Dispatching Station is located in each Region. In addition, there is a National Dispatching Station at Paris entrusted with the task of co-ordinating the activities of the eight Regional Dispatching Stations and regulating the exchanges with the neighbouring countries and the interconnected system frequency. The National Dispatching Station regulates frequency by sending control impulses to a number of generating stations spread all over the country through remote control and the Regional Dispatching Stations have no role to play in this task. With respect to the neighbouring countries, however, tie-line load bias control is adopted.

However, a restructuring of electrical regions and dispatch centres of Electricite de France (EDF) is currently in progress and it is likely that the number of regions and dispatch centres will undergo a change.

2.4.3 India

The country is divided into five electrical regions each region comprising a certain number of State systems. Each State system has or will have a main dispatch centre with one or more sub-dispatch centres located at suitable points in the system.

With the development of super-grid at 400 kV, large generating stations, and inter-regional interconnections, the Power Grid Corporation of India - a National Service - is entrusted with the development and co-ordination of exchanges of power between States in each Region and between the regions and ensuring co-ordination of operation at the National level.

The Regional Dispatch Centres have to receive their powers from the State Electricity Boards so that they could exercise their operational responsibilities and fulfil their role of co-ordination with all the necessary authority and effectiveness.

The National Dispatching Centre as it develops would prepare and co-ordinate the operations of the systems at the National level by orienting and supervising the actions of the Regional Dispatch Centres.

The operational organisation of the electrical systems in India will thus evolve towards a three-level structure of hierarchy: National, Regional and States, whose respective functions and responsibilities have to be carefully defined. For this purpose, it would be necessary to arrive at the following arrangements:

a) The generation from hydro, thermal and nuclear power stations owned by the National Hydro-electric Power Corporation (NHPC), National Thermal Power Corporation (NTPC), and Nuclear Power Corporation of India (NPCI) should be placed directly at the disposal of the PGCIL and the Regional Dispatching Centres will utilise this power to the satisfaction of the constituent States in accordance with the shares allotted to them from time to time.

b) The lines and sub-stations at 400 kV as well as certain lines and sub-stations at 220 kV selected with their importance and role in system operation should constitute the Regional Grid system and should be placed under the direct authority of the PGCIL/Regional Dispatching Centres.

c) From the operational point of view, each Region should be considered a unit. The generation from the Central power stations of NHPC, NTPC and NPCI and the generation from the State Electricity Boards will be dispatched in accordance with the generation schedules drawn up in consultation with and the approval of the constituent States. The lines between the States of the same Region will be free-flowing - the regulation of exchange being applicable only to inter-regional ties.

The responsibilities and functions at different levels have to be clearly defined. In broad terms, the National Dispatch Centre would be in direct touch with the Regional Dispatch Centres and would have no direct action on the systems, except for the regulation of international tie-lines when these are established; the Regional Dispatch Centres would have the responsibility of regulation of generation from National power stations and Boards' generation on delegation from the National Services and the Boards respectively and inter-regional exchanges, with conventions established between the Regional PGCIL and the State Electricity Boards defining clearly the distribution of tasks between the Regions and the States for ordering operations; and the State Dispatch Centres would be responsible for operations in the power stations and the networks under their charge in conformity with the guidelines formulated by the PGCIL in association with the constituents.

2.5 Location of Load Dispatch Centres

The dispatch-control centres are vital components of an utility. Therefore, they must be located at appropriate sites and equipped with the various facilities required for reliable and safe operation.

In order to carry out the functions assigned to the load dispatcher, the dispatching stations should have the facility of reliable telecommunications with all the important points of the network under his control. In fact, a reliable communication system is the backbone of dispatching and constitutes a major factor in deciding the location of a station. The load dispatcher may utilise power line carrier communication (PLCC) facilities or telephone circuits (ordinary or dedicated) or radio links. The load dispatching stations are located neither on the basis of centre of gravity of loads or generation nor on the consideration of proximity to main generating stations/sub-stations or the junction point of extra-high voltage (EHV) transmission lines. The broad principles governing the location are as follows:

a) The dispatching station should be close to an important point of the network so that PLCC facilities are easily available and communications with important points on the network are facilitated. Very often, such points are big load centres and consequently may be near a big city or industrial area. In case such points happen to be only the junction points of a number of transmission lines and are not situated near a city, they have to be kept out of consideration for locating a dispatching station for reasons mentioned in (c).

b) It should be as close as possible to the place where dependable P&T communication facilities are available and from where a few channels may be hired out for power system operation. These channels may serve as an alternative to PLCC channels or may be the only means of communication between certain points.

c) As the load dispatching situation is the nerve centre of power operations, and personal contacts between the dispatching station and the key personnel of the organisation often become essential, it should be easily accessible by road, rail and, if possible, by air.

References

1. E. Mariani and S.S. Murthy, *Control of Modern Integrated Power Systems.* Springer-Verlag, London, 1997

CHAPTER 3

FACILITIES AT LOAD DISPATCH CENTRES

3.1 Equipment and General arrangement

At the load dispatch and control centre, to accommodate the sophisticated equipment and the operating personnel, facilities have to be made available which create an atmosphere for smooth working.

The centre typically comprises:

Telephone Exchange, Telemetry and Supervisory control equipment, Display Board, Computer System, Load-Frequency control consoles, Air-conditioning plant, battery, Uninterruptible Power Supply (UPS), and associated facilities.

For housing this equipment and accommodating the System Supervisor, the support engineers and staff, one possible arrangement could be as follows:

- The ground floor provides for a Telephone exchange, Telemetry and Supervisory Control, stores, air-conditioning/heating plant and Diesel generating set.
- The first floor is the control area where the computers with their main frame and peripherals, display board, Dispatchers and Supervisory control consoles, Load-frequency control, uninterruptible power supply and battery room, are housed.

3.2 Building

The building should be designed with an appropriate electro-magnetic compatibility level (that is, protection against electro-magnetic disturbances like lightning) for protection of computers and telecommunication equipment which are sensitive to such disturbances.

The layout of a typical Load Dispatch Centre in part is shown in Figure 3.1 and 3.2.

The portion housing the control room, computers, telecommunications equipment and electrical and other building facilities has restricted access and is designed to be impregnable to civil disorders, secure from fire or flood, and self-sufficient for two weeks.

Figure 3.1: *Control room of national control center - ENEL*

3.3 Control Room

3.3.1 Mosaic Diagram

The control room houses the mosaic diagram, also called mimic board, and control desks and instrument consoles.

The Mimic Board is an indicator of the power system in the form of a single line diagram which may be depicted preferably in a semi-geographical configuration. This is made out of 15 mm x 15 mm or 25 mm x 25 mm or 50 mm x 50 mm plastic rectangular tiles, also called mosaic. These are fitted on to a grid of aluminium or any other suitable material. These tiles are engraved or painted with the symbols of stations, feeder status, line loading indicator, etc. The discrepancy switch provided to represent the breakers will have a steady lamp in the position corresponding to the breaker or switch status and will be flashing in the discrepancy portion.

The display boards subtend an angle of 110° (horizontal) and 25° (vertical) at the eye of the dispatcher.

Although alpha-numeric visual display units (VDUs) are now in general use for indication, metering and alarm display, the mimic board is used for display purposes and serves a useful purpose in that it gives the operator an immediate indication of the state of the power system.

The display of the system network is made automatically by telemeter indications. The system diagram, while providing an overall view of the configuration of the main system, is also available for emergency use, as the indications are independent of the computer.

Transmission lines are depicted by tapes colour-coded to show voltage levels.

At a strategic location on the mimic board are provided instruments giving Digital system frequency indication, System demand, Inter-system MW transfers, Synchronous time, Astronomical time and Time error.

Luminous load indicators giving the quantum and direction of power flows in the network are also provided to obtain a general idea of the situation in the network. Special signal lights are used to identify overloads.

Lamp indicators are used to depict the various devices such as circuit breakers and disconnect switches. These comprise four lamps - two red lamps to indicate the device closed, one green lamp to indicate the device open, and one white lamp to indicate that the point has been selected by the supervisory control. They are generally of the same size as the display board tiles and can be snapped into the frame in place of a tile.

3.3.2 Computerised display

When a network is very large, the wall diagram presents a major problem - it must display the complete network and also the details of each section. With the computerised display, the conditions on the whole system are held in the computer store and are continually updated. The whole network can be scanned by a rolling ball control, but only the required section of the network or the particular substation in which the Load Dispatcher is interested appears at any moment. The position of each circuit breaker, and where necessary each bus-bar selection isolator, is shown and the power flow, and if required, the reactive power flow, are shown in numerals beside each circuit.

All the information passes through the computer store and arrangements can be made to reproduce past information. This facility is very valuable for giving the conditions on the network immediately prior to any fault or disturbance.

In the absence of the wall diagram, the size of the Load Dispatch centre is also reduced.

3.3.3 Control desks and Instrument Consoles

There may be three specially designed control desks, provided with cathode ray tube (CRT) displays, of which two are used by system load dispatchers, and the third is used by the System Supervisor who exercises overall supervision and control.

These desks are equipped with telephone facilities including internal PAX (Private Automatic Exchange), meters or recorders to give information of inter-system load transfers, system total generation, system frequency, instructors by which generation instructions can be signalled to attended stations, either independently or in groups, and instruments which give information that might affect the system demand, such as temperature and day light intensity.

Alarm enunciators comprising red indicator lamps are provided on each control desk, which flash whenever grid system alarms are received in the event of lines becoming overloaded, grid voltages or system frequency exceeding prescribed limits, etc. Alarms are acknowledged by pressing the appropriate alarm button when details of the cause of the alarm are displayed.

Slide projectors are also provided to display single line diagrams and other pertinent data.

The single line diagrams are displayed automatically when a station is selected on the supervisory console. The single line is in colour, and has much more detail than the display board diagram.

3.4 Computer System

The computer system generally comprises two process control type computers connected in duplex configuration, one being the on-line master display computer and the other the stand-by computer, that is, ready at all times to take over automatically the on-line display duties. Incoming data is fed continuously to both systems by means of separate interface units with the incoming digital telemetry. With both on-line computer systems healthy, the standby computer may be used for the continuous on-line checking of grid system security by simulating the effect of credible grid system faults or for software amendment or other similar tasks.

The computer system is arranged in two physical suites spaced apart so as to minimise the risk of fire or accident damage. The two suites are however interconnected in such a manner that the peripheral devices suited in one suite are connected to the Central Processing Unit (CPU) in either its own or the opposite suite.

Other schemes with more than two computers are possible (See chapter 1 of [1]).

3.5 Teleprinter facilities

In addition to telephone and telemetering facilities, a teleprinting system with a central exchange is also provided for communication of important information, instructions, etc. to sub-control centres, important generating stations and interconnecting sub-stations. The teleprinter system is also used for communicating details of estimated demands, power transfers, scheduled generating plant, outages, etc. The teleprinters can be ranged together for communications which may have to be sent simultaneously to other control centres. Weather maps are also received from the Meteorological office in the teleprinter room through a facsimile reproducing system.

3.6 Weather Information System

System demand changes widely with changes in weather. There is a close relationship between the weather and the demand as well as stable supply of electrical energy. Weather information is generally received continuously from the Weather Bureau. However, delays often occur in the receipt of information. Power systems are therefore considering the installation of their own Weather Information System. A good example is that of Chubu Electric Power Company, Inc. (CEPCO) which has its Control Centre in Nagoya (Japan). CEPCO has more than 180 power stations, 600 substations and e.h.v. transmission lines at 500 kV, 275 kV and 154 kV, interconnected with Kansai Electric Co. on the one side and Tokyo Electric Co. on the other side.

Figure 3.2: *Control room of regional control center, Venice -ENEL*

The Weather Information System is made up of a radar site located on the top of Mount Mikuni about 30 km north-east of Nagoya, a radar data analysing and displaying system, and a cloud image receiving system picking up signals from the geostatic meteorological satellite "Himawari".

The system monitors and measures the movements of rain clouds and thunderclouds within a radius of 200 km by detecting the reflected radio waves received by the radar antenna on Mount Mikuni. Computers at the CEPCO's Load Dispatch Head Office analyse this data and give notice of the rainy areas, the anticipated amount of rainfall and the strength of thunderstorms as well as their course and velocity. Supplementing the radar data, CEPCO use the information of "Himawari". This system receives images of clouds over a wide area, establishing their distribution, course and velocity. CEPCO uses this in the preparation of demand forecasts.

The new system displays detailed information in colour in real time. This information is transmitted to the Company's branch offices, which enables effective use thereof.

CEPCO is now able to operate the system network more effectively by controlling the output of the generators and effecting changes in transmission routes in order to minimise faults and consequent outages when thunderstorms are approaching. They can also use increased water availability after heavy rainfall by controlling the water level of the dam and at the same time decrease generation from thermal plants thus lowering the overall cost of generation.

3.7 Operational data logging

3.7.1 Service logs and records

Service logs and records are kept in the Load Dispatch Centres which are used

1. for analysing faults;
2. for passing on service data from one work shift to the next;
3. for carrying out plant checks;
4. as a basis for service statistics and for planning extensions; and
5. for power billing.

3.7.2 Loading log

The Load Dispatch Stations prepare weekly and daily generation schedules for the system stipulating the half-hour MW load to be maintained at each generating station. The actual half-hour sustained MW loadings maintained at each station are logged at the Load Dispatch Station. This logging is used for constantly reviewing the scheduled generation and the actual generation at each generating station and for analysing the reasons for variations, if any.

The half-hourly units generations at each station are also logged for computing the half-hourly sustained demands in MW. The sustained loads at the various stations in the system are totalled to get the half-hourly sustained system demand. The daily load curve of the system is then drawn.

A careful study of the daily load curves of the system over a past period helps the Load Dispatcher in daily load forecasting on hour-to-hour basis and for economic allocation of generation to various sources.

The hourly MVAR loadings at each generating station are logged at the Load Dispatch Station. This logging is reviewed from time to time so that the MVAR generation is optimal and the voltage profile is satisfactory. The hourly voltage levels at major substations in the system are also logged for ensuring satisfactory voltage levels during peak load hours and the voltages are not excessively high during low load periods.

3.7.3 Measured value log

In large load dispatch stations, the measured value log comprises several current, voltage and power values which are transmitted to the control room from several system points or neighbouring control centres. These values are read off at fixed intervals (e.g. every 15 min. or half or full hour) and scanned in a rigid sequence. For the automatic log, a printed form is used in which the consecutively scanned values are entered.

One of the main reasons for transmitting measured values to the load dispatch station is to see whether they remain within the prescribed limits.

A schematic diagram of the measured value logging unit is shown in Fig. 3.3.

The arrangement shown in the figure is based on the assumption that the values are transmitted in analog form (D.C. voltages representing any electrical quantity such as current, voltage, active or reactive power, etc.), either directly from the neighbouring stations by cable or from remote stations by pulse-code signals or via remote control units.

The measured values are continuously applied to terminals 1 to n as D.C. voltages. One measured value at a time is switched on to the 'analog-to-digital converter' by means of the 'measured value selector'. In order to fully utilise the measuring range of the 'analog-to-digital converter' and thus keep its error small, the voltages fed into the analog-to-digital converter are preferably of the same magnitude. After conversion, the digital value is multiplied in the 'scale factor converter' by a factor pertaining to the measuring point. It is thus shifted back in the desired decimal range and fed into the 'measured value store'. The 'print controller' takes it from the store and passes it on to the log typewriter with time, commas and intervals added if necessary. This process repeats itself, with the 'measured value selector' traversing all the measured points step by step.

3.7.4 Switching log

The switching log is made out from signals received at the Load Dispatch Station and generally lists

- Warning and danger signals
- Relay signals for system occurrences like tripping of machines, interconnecting grid feeders, etc.
- Circuit-breaker and isolator signals
- Switching commands

In a large load dispatch station, the signal log would cover several hundreds of items of information. Each change, i.e., the arrival or dispatch of a signal or the operation of a switchgear unit, is recorded together with the time of occurrence. Where fault analysing is concerned, the successive signals initiated by protective relays are preferably recorded in the correct time sequence. Under ideal conditions, that would mean entering the signals in the correct order and dated at intervals of tenths or hundredths of a second. Most of the signals are transmitted to the load dispatch station via remote control equipment. Transmission may thus take several seconds and signals may not arrive in their correct order. A signal resolution that closely matches the ideal state can therefore be achieved only if the exact time of origin of the signal is ascertained in the remote station and added to the message.

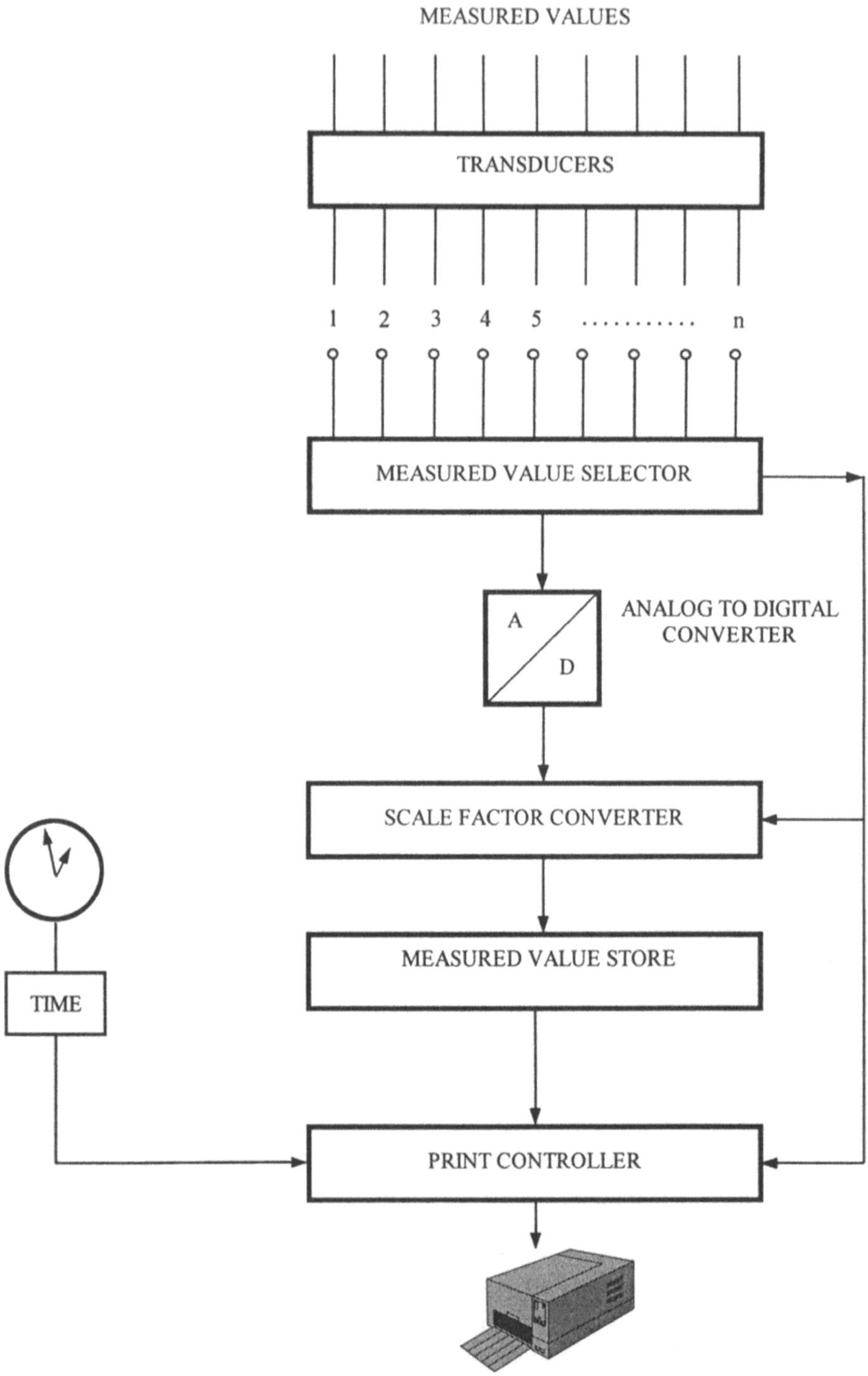

Figure 3.3: *Acquisition of local analog values*

The correct order of the signals in the log could then be re-established by a sorting process.

3.7.5 Generation plant availability

Information regarding generating units on line, units not running but that can be put on-line when required and units on maintenance and repairs together with details of repairs is collected and logged every morning. These details are also put on wall display at the Load Dispatch Station.

3.7.6 Energy generation

The energy generated at each power station in the system everyday from zero to 24 hours is logged at the Load Dispatch Station. The energy consumed by auxiliaries at each station from 0 to 24 hours each day is also logged. From the 0 to 24 hours energy generation and auxiliary consumption at all power stations in the system and also the energy import/export, the daily system consumption and the monthly and annual system consumptions are computed.

The meter log covers several energy meter readings transmitted from the power stations of the system. Depending on the particular maximum-demand period, the values are recorded as simultaneously as possible every quarter, half or full hour.

The readings are transmitted partly by telephone and partly in a coded form via remote control. The latter method is now widely used since it employs a minimum of channels and ensures reliable transmission.

3.7.7 Inter-system energy transfers

The half-hourly or hourly energy export/import from one system to other is logged at the Load Dispatch Station. From these readings the daily inter-system energy transfers are computed for raising bills.

The hourly MW/MVAR flows along inter-system tie lines are also logged.

3.7.8 Reservoir levels, inflows, etc.

The following data are logged daily in respect of each reservoir in the system:

Reservoir levels and storages at a particular hour (say, 06:00 hrs.) everyday.

Rainfall and inflows into each reservoir during the past 24 hours.

Water discharged from the reservoir for power generation during the past 24 hours.

Water surplussed, if any, through the sluices.

It is very important to have a close watch on the storages available in the various reservoirs with a view to avoiding spillages during the rainy season and premature emptying of the reservoirs during other months. A systematic study of the inflow patterns into the different reservoirs is undertaken from this logging to enable the forecasting of inflows for future periods. This exercise will enable the

Load Dispatch Station to evaluate the possible water drawals from the various reservoirs. From the level vs. storage capacity curves the storage in million cubic metres corresponding to the level and the equivalent in MU (Million of Units of energy) are determined. Once the storages and storage equivalents are determined the inflow equivalent can easily be determined as follows:

Storage at 0600 Hrs. on, say, 22nd	42.47 Mcm/100 MU
Storage at 0600 Hrs. on 21st	41.9 Mcm/98.7 MU
Units generated in the power house for the period 0600 Hrs. on 21st to 0600 Hrs. on 22nd	1.4 MU
Water consumption	1.4 x 0.4247 = 0.6 Mcm
Inflow for the 24-hour period 0600 Hrs. on 21st to 0600 hrs on 22nd	42.47 -41.9 + 0.6 = 1.17 Mcm or 100 - 98.7+ 1.4 MU = 2.7 MU

A close study of the inflow charts for each reservoir for the past years will give the load dispatcher a close pattern for any average year/good year/bad year which will be of great assistance in forecasting for future years.

In preparing generation schedules, the load dispatcher has to take into account the storages in reservoirs, inflows anticipated, irrigation commitment, if any, etc. During heavy seasonal floods, some of the small capacity forebay dams may surplus. The quantity of water surplussed during such occasion is also recorded for purposes of computing the inflows. It should be the effort of the Load Dispatcher to intelligently anticipate such seasonal inflows based on the pattern in previous years and keep the levels at the concerned forebay dam at the lowest possible level so as to receive the heavy seasonal inflows with minimum surplussing.

3.7.9 Fuel consumption and stocks

The daily fuel consumption at each thermal station in the system is logged and the fuel stocks as at the end of each day and month are computed.

3.8 Auxiliary Power Supply

For the main equipment in the load dispatch centre to function properly under all possible circumstances, the following equipment is separately installed:

- DC battery system
- AC stabilised supply
- Uninterruptible power supply with auto change-over facility
- Diesel-generator set
- HVAC system.

3.8.1 DC battery system

A special battery system is always designed to cater power supply for:

- Indicating lamps
- Discrepancy lamps
- Operating push button coils
- Communication circuits
- Control circuits

In case of AC power failure, the DC battery system can do the functions for at least 5-6 hours.

In designing the DC battery system, Ampere hours and voltage, continuous loads and intermittent loads are considered. The battery system consists of a charger which has boost as well as float charging facility. The charger panel has protections from earth fault, battery low voltage/high voltage, battery leakage, etc. Generally, two chargers are provided so that even if one of the chargers fails, the other could function and charge the batteries. Necessary audio-video signals are also made available for indicating the status of the batteries, faults, etc.

The DC batteries could be of either lead or Ni-Cd or maintenance-free.

A special room is set apart for the battery system so that the fumes emitted by the batteries do not mix with the air system of the main equipment since the fumes are corrosive in nature. The flooring of the room is always of ceramic or acid-resistant tiles, since acid or alkali reacts with the concrete or rubberised floor. The battery room is provided with exhaust fans to remove the fumes generated in the room. Similarly, the walls and the roof of the room are painted with acid-resistant paint. A wash basin with fresh water tap is provided for the personnel to wash after maintenance of the batteries and also, in the event of acid/alkali splash on any part of the body, one could immediately dilute it by washing.

A proper maintenance record and schedule has to be kept to increase the operating life of the batteries. In some countries special battery maintenance squads are deployed to look after the maintenance as well as new projects.

3.8.2 AC stabilised supply

A smooth filtered, harmonic-free, stabilised AC voltage is always desirable in the areas where electronic equipment is installed.

The AC stabiliser is designed for individual equipment or for a cluster of equipment, although the former is desired from the point of maintenance, reliability, downtime, etc. The kVA rating of the stabilisers is designed taking into account the continuous load of the operating equipment. The stabilisers contain filter circuits, buck/boost facility and also cut-out facility in case of under-voltage. The stabilisers should have a quick response time to act, so that spikes in the voltage or high current peaks do not damage the costly equipment. The stabiliser is protected by fuses, Miniature Circuit Breakers (MCBs), etc.

3.8.3 Uninterruptible Power Supply (UPS)

Power supply for the control system comprising computers and telecommunication facilities should be of the uninterruptible type (UPS), that is, the A.C. supply from the external network is converted to D.C., backed by a battery and reconverted to A.C. to feed the control system. (Fig. 3.4).

The UPS consists of an inverter, battery and battery charger. The inverter converts the D.C. supply of the battery to A.C. supply.

Loss of power or voltage transients would have detrimental effect on the computer system. If, during programming or data storage process, power disruption occurs, tremendous loss in time as well as data would occur. Therefore, every computer system is backed up by no-break UPS system.

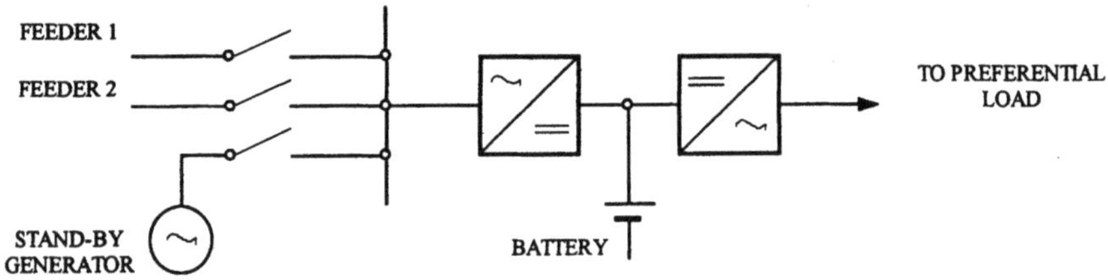

Figure 3.4: *Uninterruptible Power Supply of a Dispatching Control Centre*

The UPS system consists of dual battery charging system, with DC lead acid or Ni-Cd cells or maintenance-free batteries which in turn are connected to an inverter which converts the DC supply to either 3-phase or single phase AC supply at the desired frequency of 50 Hz or 60 Hz. The AC supply desired from the inverter is always floating across the main AC bus. In the event of the mains failure, the bus would still be energised from the UPS system without actually affecting the operation of the computer system. Also the UPS is tapped for a few emergency lights. The UPS system is generally designed for 15 to 30 minutes continuous operating rating.

All other protections, room design, maintenance schedules, etc. are totally similar to DC battery system.

3.8.4 Diesel-generator set

To make the UPS system cost effective, a Diesel generating set of equal capacity is also installed. This is provided with an auto mains failure starting arrangement. The power from the Diesel generating set is fed into the mains bus so that the load from the UPS is retrieved and transferred to the D-G set. Once the main supply is restored, the D-G set switches off automatically. During all these operations the computer system works totally uninterrupted without any loss in data or programming.

3.8.5 HVAC system

The heating, ventilation and air-conditioning system is invariably installed in a load dispatch centre to create a comfortable working environment for the personnel and a controlled climatic condition for the equipment. The temperature and humidity requirements are stringent particularly because of the computer systems involved.

The capacity of the heating and air-conditioning plant is based on the number of personnel occupying the room, density of the equipment, heat generated by the working equipment, etc. The guiding parameters are maximum temperature being 32° C and relative humidity around 50%. The temperature, even if lowered, would not have any detrimental effect on the operation of the equipment since the basic aim is to keep the equipment from getting overheated, whereas if the relative humidity is below 30%, static discharges would set in and if it increases beyond 80%, erroneous operation of magnetic tapes, etc. would occur.

The heating and air-conditioning systems are either central type or package units in sizes of 5 to 20 tonnes for the rooms where equipment is installed. However, the rooms exclusively occupied by personnel could be provided with window air-conditioners.

3.8.6 A case study

A typical arrangement of the power supply system (in a Swiss system) is shown in Fig. 3.5. The loads have been classified into non-essential (N1), essential (N2) and emergency (N3).

A) Supply without emergency power (N1):

The equipment generally covered under this category comprises:

- Yard lighting
- Lighting for rooms occupied by the personnel
- Air-conditioning provided for rooms occupied by the personnel
- Water supply pumps.

A step-down transformer with standby facility is provided to feed non-essential loads. These transformers are fed from the grid supply which is considered to have the least degree of reliability.

B) Supply with essential power (N2)

This power supply serves the equipment which may only be disconnected from the supply for brief periods. These include circuits which must be ready to operate within 5s at the most.

This system which operates with a higher degree of reliability is fed from a special transformer, with a standby transformer. If the distribution voltage drops below 70% of its nominal value (380V, 400V or 440V as the case may be) for at least 0.5s, an emergency generating set is automatically run up. The emergency set can pick up the load within 5s.

As soon as that section of the affected network can be put back into operation again, the emergency supply switches over to the normal supply with an interruption of only 0.3 s.

This takes place on the assumption that certain conditions are fulfilled: firstly, the affected network must have been in order for at least 10 minutes; the same applies to the emergency generating set, which must have been running for at least 30 min. When these conditions are fulfilled, the supply is switched back to the normal network.

These switching operations are performed and monitored automatically.

C) Supply with emergency power (N3)

This is meant for equipment which should not be interrupted under any circumstances whatsoever, such as tele-operation equipment, system control devices, data-processing equipment and lighting in the room containing the system controller.

The supply is driven by a synchronous motor mounted on the same shaft, which is fed from network N2. Normally network N3 is strictly synchronous with N2, i.e., the two are rigidly interconnected through the machines. Also mounted on the same shaft is a d.c. machine which normally generates power to charge the station battery, i.e. , when there is no fault in the h.v. network.

If the normal network fails, the a.c. supply to the above is interrupted, whereupon the d.c. machine, now acting as a motor fed by the battery, continues to keep the generator running. The drop in voltage and/or frequency is employed as the criterion for switching over to the battery. The battery can maintain full load for one hour.

However, after 5 seconds, the diesel set will restore the A.C. supply to the set. Should the diesel set fail, the supply to the auxiliaries can also be switched over to the turboset. In the rare case that the above “three-machine” set is unable to carry out its duties owing to a mechanical or electrical defect, the bypass system becomes operative and the essential loads are connected to Network N1 with an interruption of only 0.1 second.

All the control voltages required for the automatic control of the uninterruptible A.C. supply are provided by independent sources of current.

Table 1 summarises the facilities provided at a typical Area Control Centre of the Central Electricity Generating Board in the U.K.

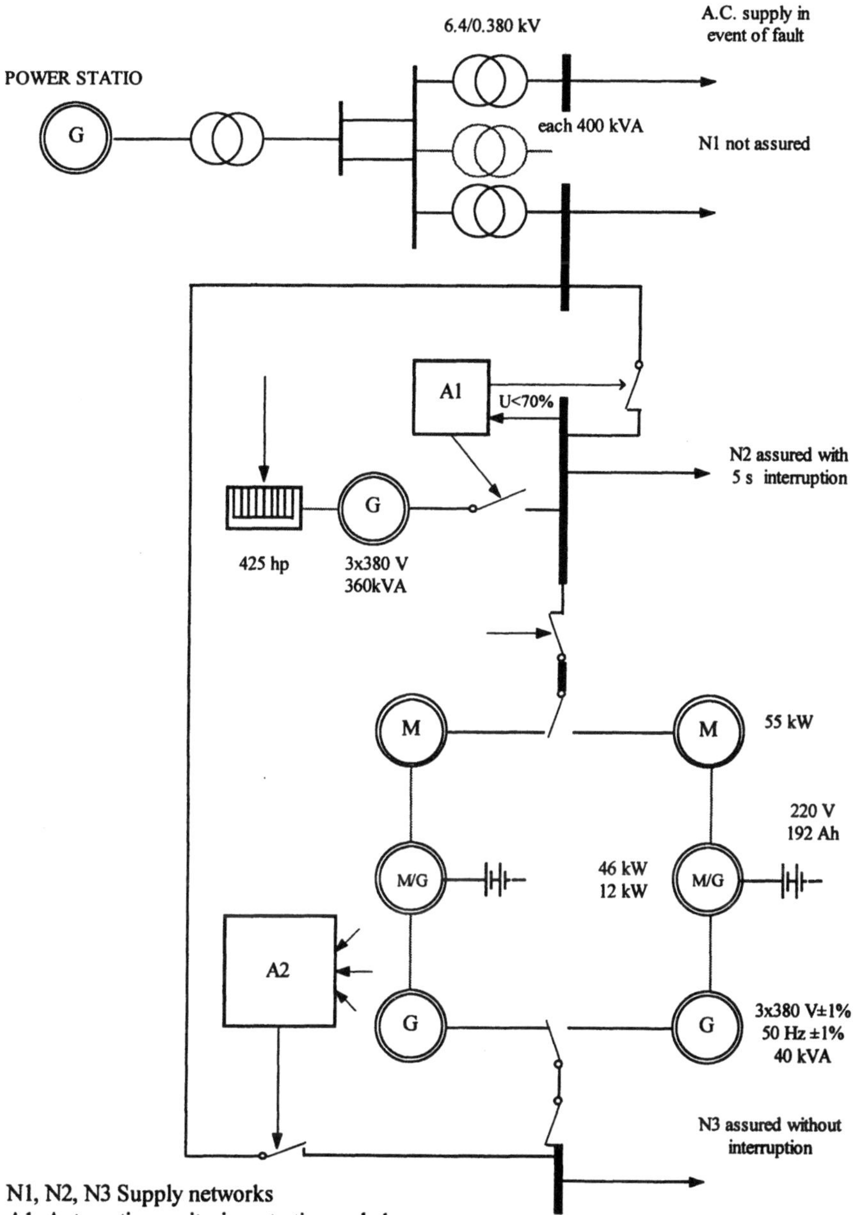

N1, N2, N3 Supply networks
A1 Automatic monitoring, starting and changeover gear
A2 Automatic monitoring and changeover gear

Figure 3.5: *Arrangement of typical auxiliary power supply system.*

3.9 Training of System Operators

3.9.1 General

One important factor influencing the reliability and security offered by the control centre is the role of the operation staff, their training, ability and discipline.

The principal personnel are the load dispatchers, computer operators, and telecontrol and telecommunication maintenance engineers.

The environment of the control centre has considerable impact on the man. In fact, the man and the environment go together. On the one hand, he has an influence on the functioning of the control system, and on the other, he is influenced in his action by the possibilities and the limits offered by the system.

It is always a good practice to associate the operation engineers and operators with the definition of the functional specifications and design of the control centre right from the beginning. They will then know the potential of the system and also its limitations.

This applies to control centres in their initial stages, but as they evolve in consonance with the growing power system and its sophistication, the staff, as and when selected, trained and put into position, grow up with the control centre and adapt themselves to it and vice versa.

There is also the possibility that, because the modern control centre is so very reliable, there are few opportunities for the operation engineer to test his abilities. For example, in Germany where the quality of power supply is very high, area black-outs occur about once in ten years. Consequently, control centre operators with a few years on job have generally no practice in restoration after large disturbances or black-outs. Even long-serving operators may have small or antiquated knowledge in these phenomena, because the performance of the network has changed in the course of years. This situation is remedied through periodical refresher training courses by way of workshops and simulator training.

Further, because of extensive man/machine communication, close co-operation and co-ordination is necessary among the different personnel involved. The computer operators execute programme running and software maintenance to suit the dispatchers requirements. The telecommunication and telecontrol maintenance staff receive all the information required for the dispatchers and have to inform them of all abnormal situations occurring in their systems. Here arise human behavioural problems like motivation, co-operation, discipline and order, and the tools necessarily are training and simulation, education and instructions.

3.9.2 Placement of staff

Training of operators is best done with reference to the particular system where they have to serve or are already serving.

Table - 1 : *Facilities at one Area Control Centre of CEGB*

			CEGB 62,888 MW	
			Main (Grid) Control Centre	Telecommand Control Centre*
Number of Control Centres (per Board)			1 National 7 Area	140
			Figures given in columns below are for the Manchester Area Control Centre	
Number of Transmission or Generating Stations (per Centre)			180	2-6
Switches (Circuit-breakers or isolators) indicated			1950	20
Alarms or auxiliary plant indicated			1005	90
Telemeasurements			517	9
Totalled measurements (Tariff and statistical)			107	4
Telecommands		(Main and Auxiliary plant)	NIL	80
3 kHz (nominal) channels	Radial	Radio	2	2
		Land-line (rented)	41	4
		Land-line (privately owned)	NIL	1
3 kHz channels	Cross-country	Powerline Carrier**	52	-
		Radio	2	2
		Land-line (rented)	123	3
		Land-line (privately owned)	10	NIL
Number of signalling channels		50 - 9600 bits/sec	780	15
Number of computer data-channels		(included in above totals)	2	NIL

*** Telecontrol is used mainly for the control of the 132 kV system and for minor 275 kV substations.**

*** * PLC is used for protection purposes only**

Note : The data pertains to the year 1976

There is considerable difference in the qualifications and background of the operators working in different organisations. For example, in most European countries and in many developing countries like India, system operation is performed by graduate engineers. On the other hand, in Japan and USA, the system operators are not graduate engineers but undergraduates.

The selection of proper personnel for operators positions is very important. In Japan, for example, the personnel are initially selected out of those having 3-5 years of experience in plant maintenance and attached to the lowest control centres. During their shift duties, they are given intensive on-the-job training and practical simulator training. Thereafter, selection is made in terms of knowledge, skills, and aptitude for posting to the higher dispatching centres. Thus, through the process of systematic training and reshuffle, well-qualified persons are selected to take up senior operating positions. Fig. 3.6 depicts this process of selection.

3.9.3 Training facilities

Training facilities are essentially intended to cope with the system disturbances as and when they arise and vary from system to system.

The disturbances are broadly categorised as: Load shedding, Black-out in one's own supply area (and normal conditions in the surrounding networks), and Black-out in the own supply area as well as in the surrounding networks.

For coping with major disturbances and network restoration after black-out, the control centres have evolved certain restoration strategies depending upon the types of disturbances.

Operators' training has been evolved in two ways:

(1) Workshops on:

a) Methods to <u>prevent</u> major disturbances

b) Methods to <u>restore</u> the network

(2) Training by simulation.

3.9.3.1 Workshops

(1) Workshops on Prevention:

These are more lecture- and theory-oriented and deal with the following topics:

- Network systems in normal and disturbed operating conditions; e.g., performances of frequency and power control equipment, operating reserve, load shedding concepts, etc.
- Power units in normal and disturbed operating conditions; e.g., stress on turbine-generator units in case of short circuits, stability limits, tripping to houseload, influence of excitation on voltage level and stability, etc.
- System performance in dynamic and abnormal conditions.

(2) Workshops on Restoration:

These are more practical and discussion-oriented and deal with:

- Restoration strategies after major disturbances or black-out.
- Discussion of real disturbances experienced in Europe; their causes and courses, with critical discussion of the restorative actions in the control centres concerned.
- Simulation of assumed large disturbances in parts of Germany; common discussion on how to recognise the state after disturbance, and search for the best restoration strategy.

3.9.3.2 Simulator Training

The purposes of simulator training are:

- Improvement of knowledge and practical skills of the individuals,
- Increasing the team-working capability under abnormal conditions,
- Development of operator's tolerance for psychological stress and capability of coping with the abnormal situations confidently,
- Establishment of an effective education and training program for the operators, and
- Improvement of the restoration procedures so that the operators can use them more effectively.

Contrary to simulators used for training aeroplane operation, and nuclear plant operation, each existing power system is unique. An ideal training facility, which is simulating to the operator exactly *his* network in exactly *his* control-room environment, will cost about as much as the one original control centre itself.

To avoid this cost/benefit divergence, attached training simulator versions utilising a maximum of hardware and software of the original control centre have been developed. On the other hand, Electricite de France and utilities in Japan install stand-alone versions in training centres far away from the control centres, and accept a high effort to model serviceable copies of their networks and control rooms.

The data volume for network, power plants, diagrams, etc. and mapping a power system usually calls for several man-years of work. For an attached training simulator version, part of the data can be obtained as a copy of the original data, but a major share to set up and parametrise the power plant models and load-models has to be added. For a stand-alone version, the whole data set has to be newly prepared, because of the varying data arrangements in different control centres.

Fig. 3.7 shows a present-day three-computer configuration in which the third computer can be used either for training or software development. In addition, this computer can substitute one of the main control centre computers, thus enhancing the system availability.

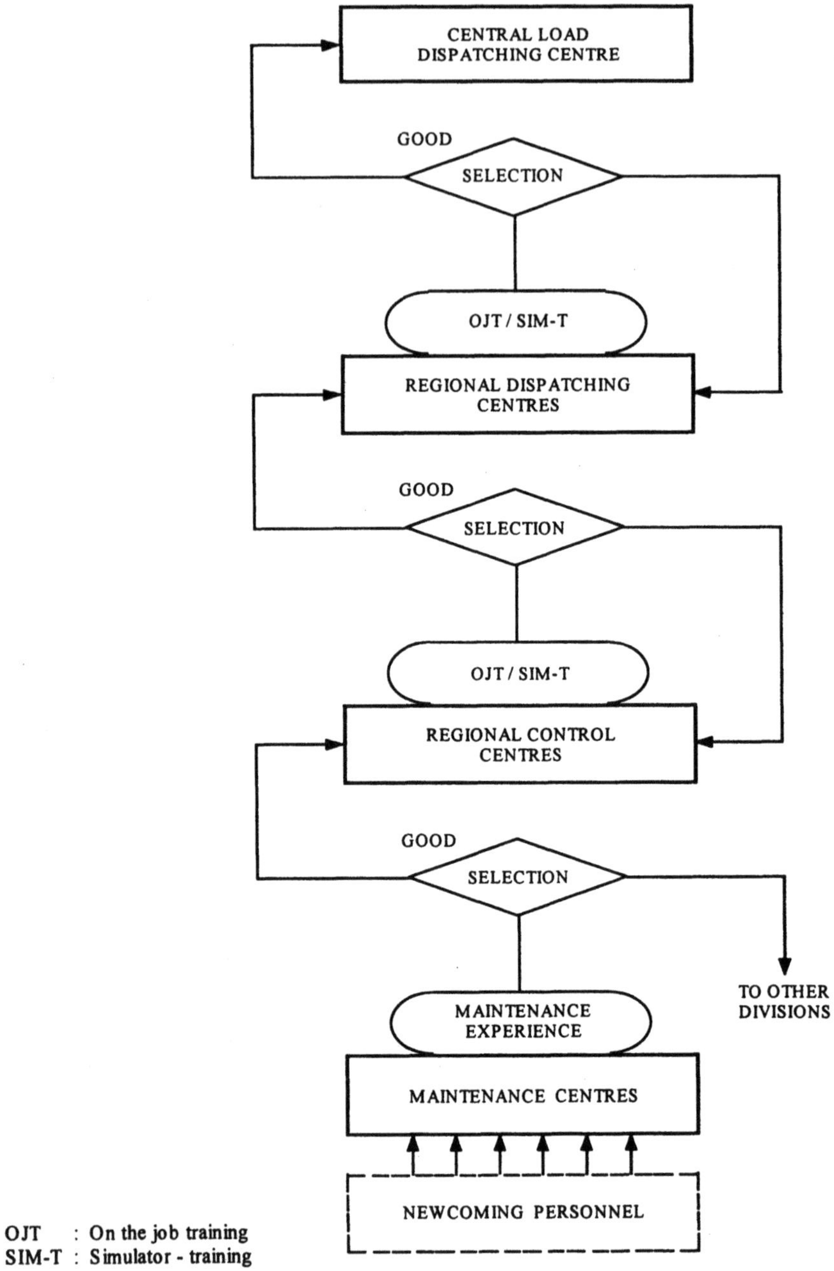

Figure 3.6: *Process of Personnel selection*

A well designed real-time *dispatchers training simulator* (DTS) is a very effective means of training system operators.

The primary objective of the DTS is to train the system dispatcher/operator to operate the power system proficiently during normal, emergency and restorative conditions. The DTS may additionally be used for other purposes such as operational planning, testing newly developed facilities, and for training of personnel other than dispatchers.

A DTS is composed of four major components - a *control subsystem*, a *telecommunications subsystem*, a *modelling subsystem* and *an instructional subsystem* (Fig. 3.8).

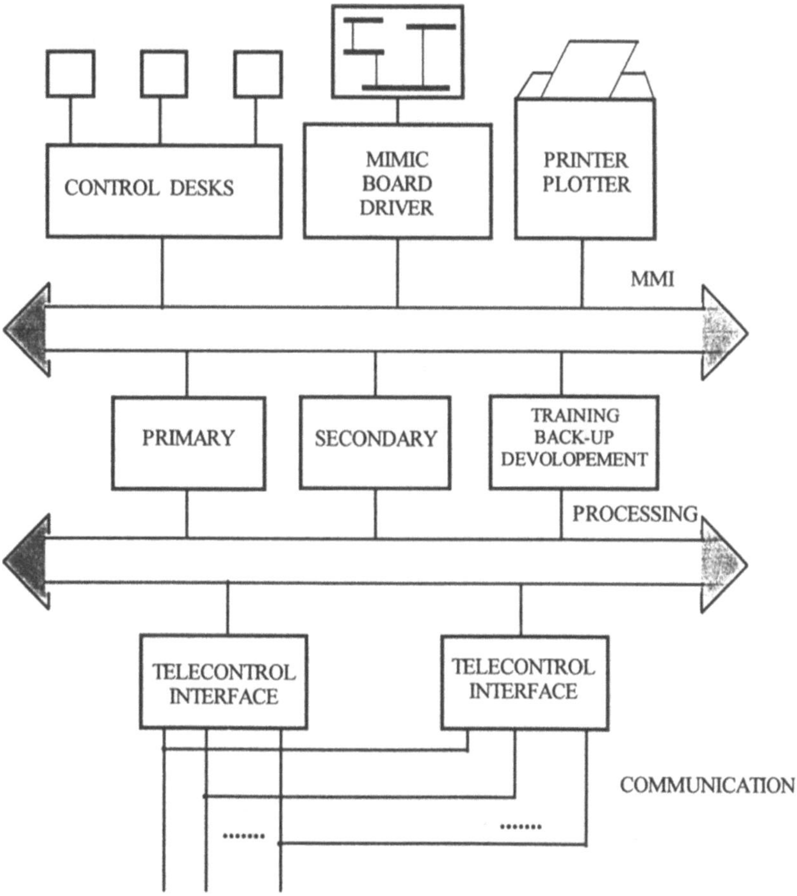

Figure 3.7: *Hardware Structure of Control Centre Systems*

The control subsystem is a one-to-one replica of the software and part of the peripheral hardware which is used for real-time power system operation. The trainee interacts with the control subsystem in the same way as he would while controlling the real power system. Therefore, the control subsystem should be a copy of the real-time control system.

The telecommunication subsystem provides an interface between the modelling subsystem and the control subsystem.

The modelling subsystem includes the models and algorithms which simulate the power system. The following functionalities can be modelled:

- characteristic features of network elements which are controllable and/or supervisable by the dispatcher;
- steady state behaviour of the power transmission system;
- dynamic behaviour of the frequency and tie-line power between control areas and the active power of the generators;
- voltage and frequency dependency of power system elements and loads;
- operation of some protection relays;
- as required, dynamic behaviour of the power system.

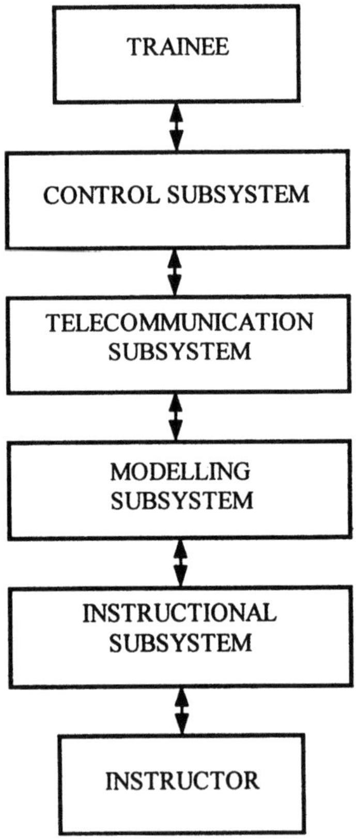

Figure 3.8: *Functional Structure of a Dispatchers Training Simulator*

The instructional subsystem includes all the tools which allow the instructor to interact with the modelling subsystem and thereby make the DTS a training tool. These features permit the instructor to change the status of the different models, e.g., switch off/on a telemetry channel, trip a circuit breaker, change the load, etc., to set up complete training scenarios and to play the role of the outside world by acting as plant operator or substation operator, or even the control centre of an interconnected utility. In addition, the instructor can observe start, stop or replay of a training session and may discuss training results with the trainee.

DTS Configurations

DTSs can be implemented in several different configurations. These can be broadly classified as:

1. stand-alone simulator;
2. simulator implemented on the back-up computer of a dual redundant Energy Management System (EMS);
3. simulator implemented on dedicated hardware, but loosely coupled with a dual redundant EMS.

A stand-alone simulator provides maximum independence from a new or an existing EMS. This means that there is complete freedom with respect to MMI, power system modelling and availability of the DTS. Thus a stand-alone DTS can be used to train power system operators for general power system behaviour using a generic power system model or for current and future power system model or for current and future power system operation of their particular power system.

However, it is very expensive to provide identical functionality and MMI when a stand-alone simulator is added to an existing EMS. Moreover, all future changes within the EMS, e.g., replacement of the EMS, changes in data base and software, MMI etc. have to be immediately reflected in a stand-alone DTS as well. This will add significantly to the maintenance costs of a stand-alone DTS.

Integration of DTS facilities with the EMS automatically provides identical functionality and MMI. In addition, data base management facilities, software and display maintenance can be common which significantly reduces support costs over the lifetime of EMS and DTS.

Historically DTS facilities provided as part of a new control centre have been implemented on the secondary computer of a dual redundant EMS in which the secondary computer is running in cold stand-by mode. With decreasing hardware costs it is becoming more and more attractive to implement the DTS on dedicated hardware loosely coupled to a dual redundant EMS. This allows combination of the advantages of a stand-alone and an integrated EMS.

Evaluation of simulator training

An interesting evaluation of the simulation training by trainees done by Chubu Electric Power Co. in Japan in 1988 is summarised below (%):

	Very much	Much	A Little	No change
1. Effectiveness of simulator training	65	30		5
2. Improvement of skills	18	78		4
3. Confidence in fault restoration	18	58	24	

The trainees response in regard to what they have acquired most through simulator training was as follows:

Confidence	..	30%
Self-improvement	..	27%
Knowledge	..	23%
Team work	..	20%

Overall the trainees recognised good effects not only in their skills for actual fault restoration but also on their psychological side.

3.9.3.3 Computer-based training (CBT)

Conventional CBT techniques have been available for over two decades, but it is only recently that the rapid development of both hardware and software has made the wider use of CBT applications possible and more desirable in different fields.

A workstation or PC is suitable as hardware. The processing and storage capabilities are rising dramatically while the overall cost is being brought down just as drastically. Interactive videodisk interfaces are available, making it possible to combine computer-generated texts and graphics with computer-controlled, random-access video and audio. The display may be on one screen or on two separate screens, with the computer graphics and text on one screen and video on the other.

The primary use of a DTS is to enhance the operator's skills for real-time and actual working conditions. The CBT system, on the other hand, extends knowledge and understanding on general theoretical subjects. It also gives the operators to a certain extent skill in controlling the individual power system, but this depends on both power system modelling and the educational program.

The maintenance of the control system and DTS with its network model is a demanding job and specialised personnel are required. Since the CBT system has no direct relation to the control system, the CBT will be more flexible in including new training subjects and changing applications than the DTS.

The investment and running costs of a DTS make it impossible for some utilities to have one. The CBT system costs are on a much smaller scale than the DTS, but the cost level will depend on the functions included and the additional systems of the CBT and most important of all, the cost of the software.

While possibilities of the CBT technique in training operator skills are limited, especially if a realistic control centre's environment is required, the CBT system can be a gradually widening training system.

References

1. E. Mariani and S.S. Murthy, *Control of Modern Integrated Power Systems.* Springer-Verlag, London, 1997
2. K. Glocker, Auxiliary Power Supply, Load Dispatching, *Brown Boveri Special Journal* (No. and year not given)
3. National Control Centre, *Central Electricity Generating Board,* December 1976.
4. J.W. Dillow, J.G. Kelly and S.A. Soutter, *Communication and Telecontrol facilities for Power System Control in Great Britain,* CIGRE 1976, paper 35-60.
5. D. Rumpel and A. Vorbach, *Studies and training simulation for power system restoration,* CIGRE 1992, paper 39-303.
6. M. Tsukade and S. Yamada, *Operator training experience by a system simulator,* CIGRE 1992, paper 39-301.
7. A. Bose et. Al., *Computer-based training for power system operators,* CIGRE 1992, paper 39-307.
8. Helmut Bollmann, *Automatic Data Logging in System and Load Dispatch Centres,* Osterreichische Zeitschrift fur Elektrizitatswirtschaft 1967.
9. S.S.Murthy, A Plan for Development of Load Dispatching Organisations in India, *Institution of Engineers (India) Journal-EL,* Vol. 50, June 1970.

CHAPTER 4
TELECOMMUNICATIONS IN POWER SYSTEM OPERATION

4.1 General

For the operation of complex power systems, particularly for load dispatching systems, it is necessary to have a highly efficient and reliable telecommunication network. Such a network has to provide facility for transmission of speech, protection relaying and control signals as well as measurands and data of all kinds.

A load dispatching centre can only perform its tasks if it is able to transmit the necessary information, to or receive it from, the transfer points or the controlling stations with utmost speed, accuracy and reliability.

Various communication media are used for this purpose: Power Line Carrier Communication (PLCC), Radio circuits, Leased Telephone circuits; in recent years Fibre Optics and Satellite communication have come into use. The choice depends on the capacity and quality required, permissible data update time, distances to be covered, the requirement of media diversity to ensure reliability and relative economics.

The communication requirements of a power system start at generating station/substation level and extend up to control centres located at suitable places in the power system. The control centres are required to interact with substations and generating stations for proper operation and control. Important parameters and events have also to be reported to the corporate head-quarters of the utility. The hierarchy of the telecommunication network should start from Remote Terminal Units (RTUs) at important remote substations/generating stations and terminate at control centres possibly via data concentrators (at nodal points of the power system in the network). From thereon, multichannel systems would be required to link them with the Area (or State) Load Dispatch Centres which in turn would be connected to the respective Central (or Regional) Load Dispatch Centres. In many systems, the Central Dispatch Centre may as well be the National Dispatch Centre as in Great Britain and France. In some systems like the Indian system, over and above the State and Regional Load Dispatch Centres, there would be a National Load Dispatch Centre (NLDC) in which case all the RLDCs would be connected to the NLDC.

4.2 Communication systems

4.2.1 Power Line Carrier Communication (PLCC)

For communication over medium and long distance power network, PLCC is the most economical and often reliable method of communication. This is essentially due to higher mechanical strength and insulation of the EHV lines which contribute to increased reliability and lower attenuation over long distances. PLCC systems support telemetry, telesignalling, telecontrol, teleprinter and teleprotection signals.

The carrier frequencies generally used are in the 50-300 kHz range and the allotment of frequencies in this range depends on the availability so as not to interfere with radio broadcasting or public telephone systems with carrier currents on overhead lines which also use the same range of frequencies. The use of lower frequency is limited by increased size and cost of coupling equipment. The higher frequency is limited by radiation losses, interference to other services and higher attenuation particularly in long lines.

The transmission conditions are not exactly the same as those applicable to an overhead telephone line. Two factors pose special problems: the variations of terminal impedance and the noise due to the statics of the high voltage lines caused by the corona along with the conductors and by the glow discharges round the insulators.

The characteristic impedance of a line between a phase and earth is about 400 ohms. The kilometric attenuation (which increases with the frequency) is of the order of 0.1 dB at 100 kHz for a 220 kV line.

The propagation laws which are relatively simple in the case of a line connecting two ends directly become complex when the lines have tee-offs. For example, for frequencies around 100 kHz., a tee-off a few kilometres long may resonate giving rise to a stationary wave system and then behave like a short-circuit.

The noise due to the statics is generally high at 220 kV and higher voltages, particularly when there is rain or snow and specially fog. In such cases, and particularly on long lines, signal/noise ratio is improved by means of compressor-expander devices.

Since telephone and data communication system with associated electronics work at low DC voltage, it cannot directly be connected to high voltage lines. Suitable coupling devices have therefore to be employed for this purpose. These consist of high voltage capacitors, line matching units for matching the impedance of the power line with that of the PLC trans-receive equipment and protection devices. Wave traps are used to prevent carrier current from entering the power equipment in the sub-station. The wave trap also reduces cross talk with other PLC circuits connected to the same station.

Side bands

On networks up to 132 kV, the carrier wave is generally transmitted with two side bands, which enables the use of very simple equipment but occupies a comparatively broad frequency band (2 x 6 kHz.). For longer transmission and above 132 kV, it is preferable to increase the signal/noise ratio on the one hand and reduce the frequency band on the other. For this purpose, equipment with a single side band including high quality filters is used. One frequency band of 4 kHz. would then be sufficient.

The 4-wire circuits so formed have the conventional audio band width of 300 to 3400 c/s. Speech with good intelligibility can be transmitted over a band width of 300-2000/2400 Hz., the remainders of the bandwidth being normally utilised for data transmission, teleprinting and control purposes. Data channels at 600 Baud have been successfully multiplexed with speech in PLC channels. As the number of line sections increases, the noise level increases rapidly, so much so that 4-5 sections generally become the limit.

Mode of coupling

For telephony and data transmission, phase-to-earth coupling (Fig.4.1) is generally used for the sake of economy, but for transmission of protection signals, interphase coupling which gives a much larger security (better propagation and functioning in the case of phase-to-earth faults) is used.

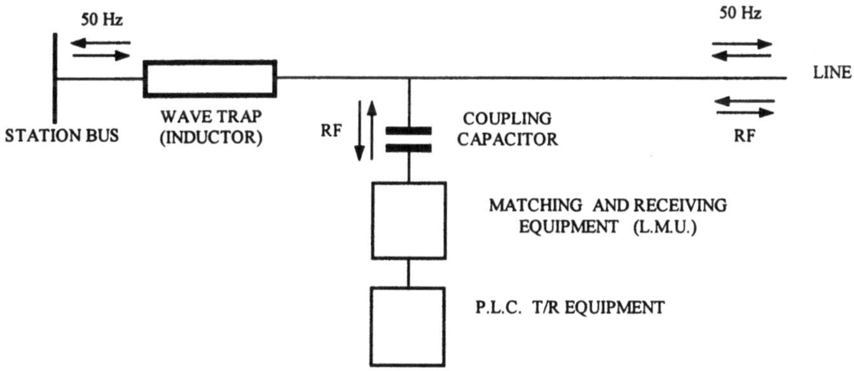

P.L.C. - power line carrier

Figure 4.1: *Basic coupling arrangement (Phase to ground)*

Equipment

The equipment is now entirely transistorised, the maximum power used being about 10 Watts and is fed by 48-Volt batteries.

There are, however, some disadvantages with PLCC. These are: PLC links are susceptible to background noise level caused by corona effect and also to impulse noise due to the operation of isolators and circuit breakers; they are sensitive to weather conditions; and the available frequency band is generally crowded making it difficult to find unallocated frequencies for new lines.

4.2.2 Microwave Communication

Microwaves as a communication medium for power systems offer a number of distinct advantages. The most significant amongst these are the channel reliability and the band width available in each of the allocated bands. The latter advantage makes it possible to use multiplexing systems which permit a single microwave carrier to be modulated simultaneously by several independent voice channels or a combination of voice channels and control of telemetering channels. The frequency bands in which microwave radio links operate are controlled by national bodies who are responsible for the allocation of the radio spectrum within their individual territories. In India, for example, the Wireless Planning and Co-ordination (WPC) Wing plans and assigns the frequency bands to different users keeping in view the broad guidelines recommended by International Telecom Union (ITU) and World Administration Radio Conference (WARC). WPC has assigned 2.3-2.5 GHz and 8.2-8.5 GHz bands to the power sector for planning narrow band networks over large areas. Microwave systems generally consist of transmitters which operate in a range between a few hundred milliwatts up to 5 watts of transmitter power, and usually have parabolic dish antennas which focus the microwave energy into a narrow beam directed towards the receivers.

The path of a radio beam in microwave frequencies is considered to be line-of-sight, which characterises a direct, open air transmission path that is free from obstructions such as buildings, trees and other objects. It is often thought of as a straight line between the transmitter and receiver. In actual practice, however, microwave signals follow neither a straight line nor a straight path but use the refractive nature of the atmosphere for establishing communication. The curvature of the earth and the slight bending of microwaves as they are refracted downward by the earth's atmosphere are two factors that must be considered while constructing microwave path profiles.

The microwave path profile is calculated showing the elevations of the path ends as well as elevations of points along the path; both the earth curvature and the desired Fresnel clearance are calculated and added in order to plot the clearance target above each obstruction (Fig. 4.2 and Fig. 4.3).

Microwave signals may still be subject to variation due to an effect called multi-path fading, popularly known as Raleigh fading, which occurs when the microwave energy is diffracted or scattered along the path so that part of the signal arrives at the receiving antenna out of phase with the desired signal. To avoid the fade margin effect, space diversity, frequency diversity or loop protection configuration is used at the receiving end.

A necessary feature in microwave systems of any size is the use of repeaters. In a power system utility, if it is possible to locate microwave system terminals at substations and power plant sites, then the site can also serve as repeater to the next station. Otherwise sites must be designed to serve solely as repeaters, which then adds to the cost of obtaining right-of-way, building access roads, houses, towers, power supplies, etc. Passive or billboard repeaters are sometimes used to reroute the signal around an obstructing mountain. Passive repeaters do not require power or expensive repeaters.

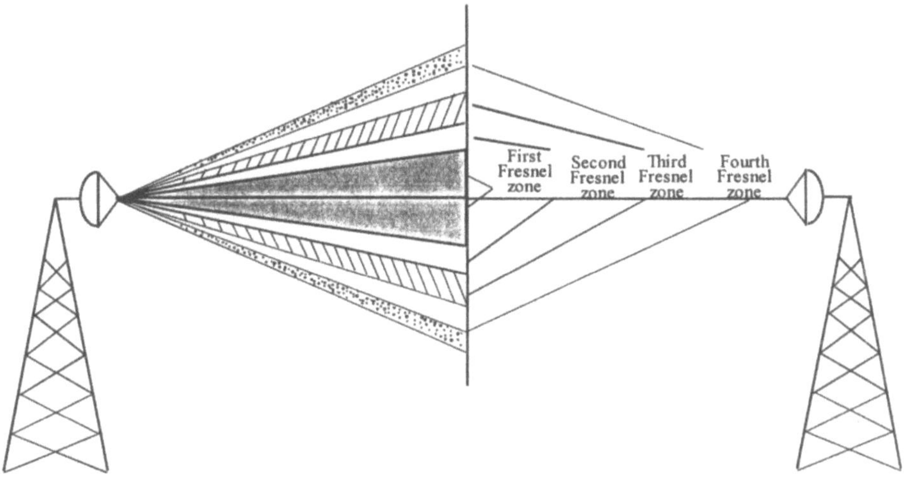

Figure 4.2: *Wave front cross-section*

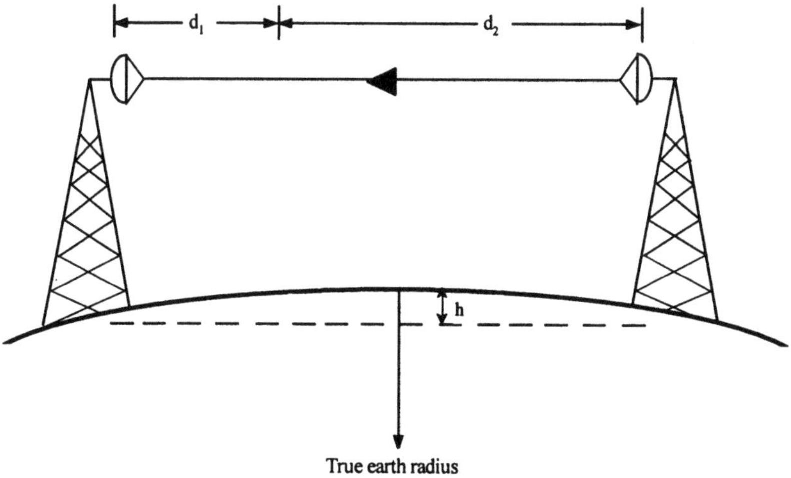

Figure 4.3: *Effect of Earth curvature on path clearance*

Microwave systems have the following distinct advantages:

- Bandwidth is sufficient to meet the needs of an electric power utility.

- Microwave radio is more reliable and secure from electric power system faults, when compared to power line carrier. Availability of radio-links is not affected during maintenance of and faults on the power lines.
- Additional services can be provided relatively easily and at low cost, in that further channels can often be added later.
- Microwave radio provides high reliability, in a hot-standby, frequency diverse or loop configuration.

Some disadvantages are:

- Repeater or terminal sites must be located every 50-60 km requiring additional land and buildings.
- Users channels are located at sub-stations which are not favourable from a radio path view-point. The substations are more usually located at low altitudes while radio links are more easily constructed between higher altitude locations.
- Propagation problems including path loss and possible fading effects.

4.2.3 Leased Telephone Circuits

Circuits can be leased by the power utility from the Telecom authorities for an initial connection fee and an annual rental charge. Leased circuits simplify the problems of developing a dedicated communication network, and the need for licensing, land acquisition, right of way, building, tower, power source, etc. is eliminated. It should however be ensured that the response time for correction of malfunctioning of communication channels is the minimum possible.

The telecom network generally comprises underground cables, overhead lines and radio links. The cables and the radio links offer a good quality of service. The Telecom services usually lease out circuits to utilities in the form of 4-wire circuits without hybrid connections or signalling equipment. These circuits usually present the conventional band width of 300-400 c/s as defined by the C.C.I.T.T (Comité Consultatif International Télégraphique et Téléphonique) and hybrid connections and signalling equipment are provided by the utility and kept in its premises.

4.2.4 Fibre Optic Communication

4.2.4.1 Optical Fibres

In fibre optic communication light is used as a carrier. Light is an electromagnetic wave of frequency 10^{14} Hz. which promises a tremendously wide band width for transmitting information compared with conventional systems.

The optical fibre is a cylindrical structure with a central region, called the core, and a surrounding external cylinder, called the cladding, with the refractive index of the core higher than that of the cladding. If now a light ray is incident at the interface between the core and the cladding at an angle greater than an angle called the critical angle, the light will propagate along the fibre through a series of consecutive total reflections.

Fibre optic cable systems used for communication are made up of the following main components (Fig. 4.4) :

1) An optical source
2) A means of modulating the light with the signal to be transmitted
3) A transmission medium
4) A photo-detector which converts the received optical power back into an electrical signal
5) Electronic amplification and signal processing required to recover the signal and present it in a form suitable for use.

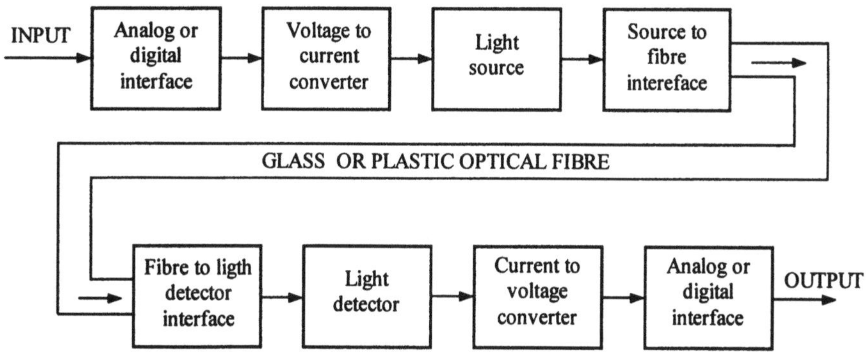

Figure 4.4: *Fibre optic communications link*

4.2.4.2 Numerical Aperture

Numerical Aperture (NA) signifies the ability to collect light from an optical source. Rays of light striking the end of the fibre are efficiently transmitted through the fibre only if their angle of incidence is within a specified number of degrees of the fibre axis. Typical values for the NA of a telecommunications type of fibre range from 0.16 to 0.25 (sin θ), corresponding to the maximum acceptance angle of 9° and 14° with respect to the axis (Fig. 4.5).

Optical fibre is made of silica which is the most transparent material. The core is usually doped with germanium and possibly phosphorous pentaoxide to enhance its refractive index. However, a fibre does have losses, and hence the signal travelling through it gradually becomes faint (attenuated), and eventually a repeater is needed to boost and regenerate the signal.

Spreading of light pulses takes place as they travel down the fibre. Due to the spreading effect, pulses tend to overlap, making them unreadable by the receiver. Dispersion limits the bandwidth or information carrying capacity of a fibre.

The capacity of a communications system is directly proportional to its bandwidth: the higher the bandwidth, the greater is the information carrying capacity. Bandwidth is enormous in fibre optic cables and could meet very satisfactorily the power utility's demand for telecommunication. In a typical 140Mb/s fibre optic system, 1920 voice channels can be transmitted simultaneously down a single optical fibre.

In fact, current limitations on transmission bandwidth in optical communications equipment are due to the switching speed maximums in network equipment and not the capacity of the optical fibre itself. Newer techniques like Wavelength Division Multiplexing (WDM) promise even higher bandwidths.

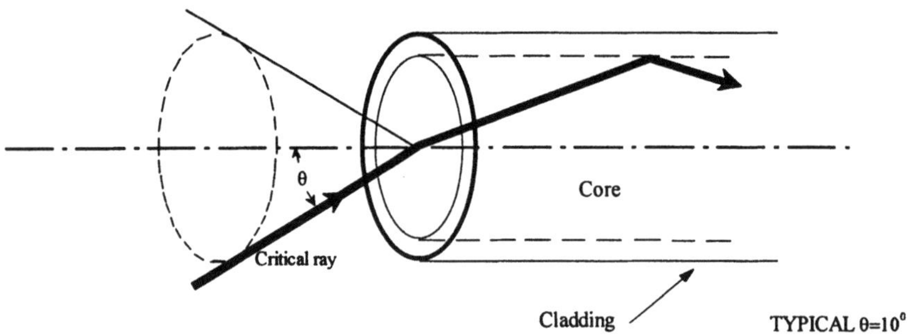

Figure 4.5: *Numerical aperture*

4.2.4.3 Techniques of Installing Fibre Optic Cable

There are three basic techniques for installing a fibre optic cable on an overhead power line: one, the optical fibre is enclosed within the ground wire to form a composite fibre optic ground wire; two, the fibre optic cable is fastened to or wrapped around the overhead ground wire; three, the cable is installed low on the tower structure with small sags so that the ground clearance may be maintained and at the same time clashing with conductors is avoided. The field conditions and relative economics have to be properly evaluated in making a choice of the technique to be employed. However, the composite ground wire system is the most preferred one particularly for new lines.

In summary, the fibre optic cable system holds considerable promise for the power utilities in view of its very high channel capacity, total immunity to electromagnetic interference that is present in substations and power plant switchyard environments, high data security and very low transmission error rate.

Electric power companies in Japan had pioneered the use of fibre optics in their operations. In fact, in 1975, a working group involving 12 power companies was set up within the Technical Development Committee of the Central Electric Board Council which conducted extensive studies with the specific object of establishing optical communication for use in power systems.

The type of applications which have been realised with optical fibre is given in Table 4.1

Table 4.1- Principal applications of optical fibre communication systems

No.	Use	Location
1	Power system protection	Between power stations/sub-stations
2	Supervision and control	Between load dispatching centre and power stations/sub-stations
3	Inter-computers	Between head office and branch offices
4	General communications	As required

Some of the typical commercial applications made in Japan are given in Table 4.2.

Table 4.2 - Some typical commercial fibre-optics systems for power application in Japan

Sl. No.	Company	Site of actual Usability	Kind of actual usability application	Route	Beginning of commercial service	Optical fibre cable Distance (km)	Aerial under-ground	No. Of cores
1	Tohoku	Sennan (C/C) to	B,D	a	2/82	15	a	2
2	Tokyo	Head Office to Ikegami (C/C)	A,B,D	f	3/82	15	b	4
3	Tokyo	Shin Noda (S/S)	B,C	b	5/84	2	b	8
4	Chubu	Tsu (B/O) to Horiki (C/C)	B	h	8/81	45	a	1
5	Chubu	Head Office to Okazaki (B/O)	D	e	9/82	61	a,b	4
6	Chubu	Shin Hokushin (S/S) to Yoshida (C/C)	B	a	3/83	15	a	4

(1) Application
- A) For protection of system
- B) For supervision and control
- C) For linkage with computers
- D) For general communication

(2) Route
- a) Between electric stations
- b) Within premises of electric station
- e) Between head office and branch
- f) Between head office and electric station
- h) Miscellaneous places

(3) Site
- S/S: Substation
- C/C: Control centre
- B/O: Branch office

(4) Fibre Cable
- Aerial :a
- Under-ground :b

Fibre-optics have also been used in the terminal stations of high voltage DC links where control and monitory signals are carried to and from the thyristor valves. Some examples are back-to-back converter stations at Acaray in Paraguay, Durnroher in Austria and on both sides of the new cross-channel link (between France and the United Kingdom).

4.2.5 Satellite communication

Communication between two stations can be established via an artificial satellite. A communication satellite located at 36,000 km or so above the equator of the earth acts as an active repeater. At this distance the satellite orbits around the earth once in 24 hours and hence it appears stationary with reference to a particular place on the earth. These satellites are called Geo-stationary Satellites.

The modulated signals are sent from the transmitter at the earth station on an uplink frequency in 6 GHz band to the satellite, where the signals are regenerated and beamed back to earth in 4 GHz band. The signals are received at the destination station on earth by means of a parabolic or spherical dish-antenna which is precisely aligned to face the satellite.

The INSAT-II series satellites in India are working in two frequency bands: C band and Ext. C band as under:

	Up Link	Down Link
C band	5.85 to 6.425 GHz	3.625 to 4.2 GHz
Ext. C band	6.7 to 7.025 GHz	4.5 to 4.8 GHz

The frequencies are divided into a number of 36 MHz bandwidth transponders. Thus there are 12 transponders in C band and 6 transponders in Ext. C band. The transmitted signal occupies one of these transponders. Depending upon the signal bandwidth and the earth station characteristics, each transponder is shared by a number of earth stations.

Very Small Aperture Terminals (VSAT) are low cost terminals to establish communication via satellite with a central location. The nomenclature 'aperture' is used synonymously with the dish diameter, which is 0.6 to 2.4 m for VSAT. VSAT consists of a transreceiver, a modem and a controller. VSAT communicates with a large earth terminal called a hub. Direct VSAT-to-VSAT connectivity is not usually feasible. Hub controls and monitors the function of a number of VSATs. VSAT and hub can be located anywhere within the illuminating area of the satellite (within India for INSAT series satellite).

There are different types of VSATs in use for different services such as telephone and low speed data, multiple telephones and high speed data, high speed data network, etc.

There are two aspects with reference to sharing of satellite's capacity. The first is multiplexing and involves the same schemes of frequency and time division multiplexing. If the Satellite were merely to provide a point-to-point link between two earth stations, multiplexing would be all that was needed. However, one of the most valuable aspects of the satellite is that it can be shared by many antennas scattered across the footprint of the satellite. In addition to basic multiplexing a scheme therefore must be devised which will permit many geographically dispersed earth stations to share the satellite. This is referred to as multiple access.

Advancement in integrated circuit (IC) technology has made it possible to manufacture compact Time Division Multiple Access (TDMA) VSAT systems which can provide advanced digital voice, high speed data and two-way video conferencing services.

VSAT technologies show very reliable operation with only basic maintenance requirements. VSATs are easy to upgrade - bandwidth can be increased from 1200 bps to 2400 bps, etc. They are relatively easy and quick to install.

There are however some disadvantages which have to be overcome. These are:

- The delay that occurs in the signal travelling far into space and back. The signal propagation time is about 270 ms. and varies slightly with the location of the earth station. The telephone user has therefore to wait for an extra 540 milliseconds for the reply to reach him if the call goes via satellite in both directions. In interactive data transmission via satellite, a terminal user will experience a constant increase in response time of about 540 ms. The system designer has to take this into consideration in designing the overall system response time.
- Since satellite and terrestrial microwave links use the same frequencies, there is a serious problem of radio interference.
- The satellite is prone to two types of eclipses. One, when the earth shadow passes across the satellite, its solar batteries stop operating. These eclipses last from a few minutes to slightly more than an hour on 43 consecutive nights in springs and fall; 277 days of the year are free from eclipses. Two, less commonly, the moon's shadow passes across the satellite, like a solar eclipse on earth. A more serious form of outage occurs when the sun passes directly behind the satellite. The sun, being of such a high temperature, is an extremely powerful noise source and so blots out transmission from the satellite. This outage lasts for about 10 minutes on 5 consecutive days twice a year. The only way to achieve continuous transmission is to have two satellites and switch channel to the non-eclipsed one before the eclipse begins.

The feasibility of using satellite data links for power system monitoring and control was first demonstrated in mid eighties in the US where control signals were sent from the West coast to a substation in New York via satellite. Since then, there has been a rapid increase in the number of power utilities using satellites for communication.

4.3 Practices in some countries

A study of the practices relating to the communication channels for power systems gives the following picture.

Canada

Hydro Quebec have utilised their own Microwave communication system for protection and control of high voltage systems up to 735 kV. They are now in the process of installing fibre optic cable all over their system.

Belgium

The electric supply companies have their own microwave links for control of high-voltage systems. They operate in the range of 410 to 465 MHz. 24- and 60-channel groups are both used with the latter forming the majority. The utilisation of channels is about 30% for telephony, 44% for data transmission, 19% for distance protection and 7% for interswitch board telephony.

Italy

There are radio links working in parallel with PLCC, thus ensuring good availability. A tropospheric radio link for special remote control has been operating between Sardinia and Tuscany since 1969. This link is an important part of the control loop of the 200 kV D.C. system between Sardinia and the Continent covering a distance of 330 km (including 300 km under the sea) and operate at 450 MHz with good availability.

Italy ranks amongst the first 2-3 countries in the world using the maximum number of PLCC circuits.

Brazil

Brazil has microwave system operating in the 7 GHz band and has about 60 links on a total route length of 2530 km. UHF system is also used for places with smaller channel demand and there are 34 links connected to the microwave routes covering a total length of 1214 km.

Great Britain

Radio channels are used by the South of Scotland Electricity Board and the North of Scotland Hydro Electric Board. The latter has an extensive 1.5 GHz microwave network which is used as the principal communication medium. The CEGB of U.K. used mainly rented telephone circuits since distances are short. The

P.O. network is extensive, wholly underground and highly reliable. Some private pilot cables are used mostly around the big cities.

In all the three Boards, PLCC is used exclusively for protection signalling.

Switzerland

The PLC network has been built up which, as far as the number of channels is concerned, has practically no rival in any other country in the world, in relation to the total length of power lines and the area supplied . With few exceptions, the entire network is equipped with multi-purpose single-sideband sets which, in addition to a telephony channel with automatic trunk dialling, carry an average of four superposed channels for protection relaying, as well as telemetering, remote control and regulation signals.

France

In the French power system, about 60% of communication channels are Post Office circuits and the remaining 40% PLCC circuits. There are also telephone lines mounted on the same towers as high voltage transmission lines and Radio communication circuits, but these are limited.

Reliability and economic considerations decide which of these means should be chosen in a particular case. In general, PLCC is used for distances of 20 to 480 km, and telephone pairs rented in underground Post Office lines are used for distances between 640 and 960 km.

The following three types of Post Office networks are generally utilised:

1) The 'administrative' network connecting the head-quarters to the various operations units. It may be used for any kind of communication: technical, administrative or commercial.
2) The 'safety' network connecting the control centres, generating stations and substations. Under the national regulations, the 'safety' network is considered as a service and the tariff is concessional. The use of this network is highly restricted to the operational safety and cannot be used for any administrative or commercial purpose.
   All the circuits of this network are connected to automatic exchanges, so that the operation of the network is entirely automatic. The equipment automatically selects the best route among those which are available between the two points to be connected.
3) The ordinary Post Office network utilised on subscriber service basis.

The control centres are provided with both 'administrative' network and 'safety' network.

India

The carrier frequencies in the range of 50-500 kHz. have been widely used in the power systems since early 1950's. However, frequency range of 150-285 kHz is used extensively for low frequency broadcasting.

The frequency band of 50-150 kHz is now available to the power sector without any restrictions subject to the following conditions:

1) The power level should be restricted to 40 Watts maximum; 20 Watts power level may be permitted for long distance 220 kV lines (exceeding 200km).

2) The PLCC system should employ phase to phase or inter-circuit coupling only.

Leasing of channels from the Telecom. Department between metropolitan cities and some large cities is possible only from the proposed National Load Dispatch Centre and the Regional Load Dispatch Centres down to the level of State Load Dispatch Centres (SLDC).

The power sector could establish its own independent system of communication below the SLDC level. This could be a terrestrial network or a satellite-based system or a combination of them depending upon the geographical nature of location.

4.4 Role of communications in load dispatch centres

The communication facilities required for satisfactory functioning of load dispatch centres have to provide for Telephony, Telemetering, Telesignalling, Teleprinting, and Remote Control.

The communication channels must be reliable and available at all times. For this purpose it has become a standard practice to provide for duplicate channels - one main and the other as backup on alternative media, with proper switching facilities.

With the advent of digital communication, the communication technology and the computer technology are both becoming inter-dependent developments today. Digital tele-communication basically implies that all information, whether voice, text or data, can be transmitted over the same medium in identical fashion, that is, in digital form.

The communication circuits have a band width of 300-3400 Hz, the lower part of the spectrum 300-2000 Hz being reserved for telephony and the upper part 2300-3400 Hz for transmission of coded information (teleprinting, telemetering, remote control, etc.). This coded information uses channels, the central frequencies of which are spaced 120 Hz apart. This separation between the telephone channel and the information channel is released with low-pass, high-pass and band-pass filters. On a normal circuit, one telephony channel and 6-7 independent data channels can be simultaneously transmitted. The different levels of power which is injected on each of these channels are standardised in conformity with C.C.I.T.T. regulations.

4.4.1 Teleprinting

The conventional teleprinters working at 50 baud are connected with automatic switches. In order that the users do not happen to find the circuit continuously busy, the auto-switches are provided with memory devices which are capable of registering a certain number of messages. This way a message is always accepted by the sending-end teleprinters. When the circuit and the addressee are free, the message is normally transmitted. When the circuit is busy, the message is registered in the memory of the auto-switch and automatically transmitted as soon as the circuit becomes free. The teleprinters which are used by the grid control centres are used for the transmission of digital data intended to be introduced directly into computing machines. The transmissions have therefore to be free from errors. For this purpose, the teleprinters are provided with error detection devices which prevent the receipt of a message when the code has been altered during the communication.

4.4.2 Telesignalling and Remote Control

The function of telesignalling is to *see* what is happening at a distance and the function of remote control is to *act* at a distance. Telesignalling and telecontrol are generally grouped together as they function on the same technique.

Remote control is always based on the principle of a relay which, on receiving an electric pulse, causes the apparatus (fed by a local power source) to function.

When the distance between the sending end and the receiving end is short, one wire may be used for each function to be fulfilled. It may then be worked by direct current, and a comparatively simple combination of relays would suffice.

When the distance is long and there are a number of remote controls, a distinction has to be made between the controls, either by combinations of several parameters (voltages, polarities, frequencies) or by a code similar to the telegraphic code.

The codes used are of the binary type and the number of digits depends on the number of equipments to be remote controlled. The transmission is effected in narrow bands, with frequency modulation, at velocities of 50 or 75 baud. On such a channel, several equipments could be remote controlled, the complete operation of remote control return signalling being effected in less than 3 seconds.

In many cases, when the equipment is merely telesignalled without being remote controlled, the principle is exactly the same for telesignalling as for remote control. The signalling corresponding to the change of position of an equipment is characterised by a more or less complex code which is transmitted to the control centre, where it causes the flashing of a lamp on the synoptical board. The system includes the same securities as for remote controls.

With the development of cyclic digital telemetering, telesignals are also transmitted likewise with as great a security as with conventional equipment. In regard to remote control, however, since utmost reliability is called for, the change-over from electromagnetic relays to semi-conductors is necessarily gradual.

4.4.3 Telemetering

At the control centre a large volume of information pertaining to the system is required and the telemetering has generally to cover:

- Active and reactive power outputs of power stations, either summated or for each unit, depending on the importance
- Exchange of active and reactive power and energy with the neighbouring systems
- Power flow on important transformers and transmission lines
- Status of circuit breakers of major tie lines and generators
- Voltage at important nodes
- Frequency at transfer sub-stations

The data plan thus prepared directly gives the volume of data required to be telemetered and this indicates the capacity of the master and remote telemetry equipment required. The data plan also indicates the data update times. Normally for control applications such as load frequency control (LFC), the necessary data like tie line flows and frequency must be updated fast -- every 2 secs. Other data may be updated every 5 or 10 secs depending upon operational requirements. The consideration of update time is very important and critical, as this very factor determines the cycle time and the related data transmission speeds. Fast update time requirements would call for higher transmission speeds, which ultimately reflects upon the total channel requirements of the whole system. For example, if transmission speeds of 1200 Bds are required, a full channel of 0-4 kHz nominal bandwidth may have to be set apart exclusively for this purpose only. However, for lower data speeds of 50 to 600 Bds, the data channel could be multiplexed with speech in the same channel of nominal bandwidth of 0-4 kHz.

In selecting any telemetry/telecontrol system, certain factors should be kept in mind: expandability and flexibility of the system, system redundancy, ease of maintenance and calibration and compatibility of interface with other systems like the computer system and the mimic diagram.

The present-day systems are fully digital and solid state employing integrated circuits and are of compact modular design leading to ease in operation and maintenance as well as great flexibility in adapting the system to the changing requirements of the user. The typical mean time between failures (MTBF) and the mean time to repair (MTTR) figures for modern telemetry equipment are of the order 18,000-20,000 hr. and 1 hr. respectively. The micro-processor based systems also offer a possibility of limit checking of data at remote stations before transmission to master station. Thus, it would not be necessary to transmit a particular measurand or a group of measurands unless there has been a significant change in its value compared to that transmitted in previous cycle. Such limits could be set and programmed as required. This would result in reducing the load on telemetry cycle and consequently on communication channels.

There are two types of configuration of telemetry and telecontrol systems from which the choice is made depending upon the functional requirements:

1) Systems based on master-slave call/reply concept, where the master station interrogates each slave station to ask for the telemetered information, in a fixed sequence.

2) Systems based on digital cyclic continuous mode of operation where the remote stations report independently of the master stations.

The telemetry systems working in many of the grid systems are of the second type. These systems are simpler but require independent telemetry transmitters and receivers for each data link, thus not offering much flexibility in the sense that for each new future link, separate telemetry system would be necessary. Telecontrol facilities are also normally not built in, as channels in command direction are generally not required in such systems. The call/reply type of systems where the master station handles a number of remote stations, offers greater flexibility in that future remote stations could be easily added to the system without any major modification. However, in the case of failure of the master station, the whole telemetry system is likely to get affected if the master station equipment is not duplicated. Thus, to achieve a high degree of reliability and availability, it has become the normal practice to plan for 100% redundancy by providing a hot standby for the master station telemetry equipment. Modules and components such as input/output devices whose failure results in loss of one or two measurands only but does not result in the failure of the complete system, are generally not duplicated.

It is equally important to consider the aspects of interfacing the telemetry system with the dispatch centre main computer system as well as with the mimic system. The interface equipment should be capable of allowing transfer of data and commands between the computer system and the telemetry system without impairing the overall system performance. The telemetry outputs should also be properly interfaced with the mimic system which includes digital displays, discrepancy lamps, line loading indicators, etc.

4.5 Telemetering Systems

There are basically two types of telemetering system--Analog and digital.

4.5.1 Analog Telemetering

4.5.1.1 Continuous Telemetering

The most common application of analog equipment in the electric utility industry is to read interchange megawatts in the dispatching office. It provides a continuous net interchange MW signal for automatic load frequency control equipment.

As shown in Fig 4.6, the transducer converts the MW into a D.C. signal, which varies proportionately with the tie-line MW power flow. The analog transmitter converts the D.C. input signal into a variable frequency output signal, which may vary from 10 to 30 Hz proportionately to the input signal. This variable frequency is transmitted to the analog receiver at the other end of the system. The transmitter is called analog because its output is analogous to the input. All analog devices work with magnitudes.

The telemeter receiver converts the varying frequency signal back to a D.C. signal which can then operate a suitable end device such as a recorder or an indicator.

In the analog system, there is continuity of the telemetering signal. In Fig 4.6, a frequency between 10 to 30 Hz is continuously going from the transmitter to the receiver. Hence any momentary channel noise or disturbance will simply result in a sharp spike on the recorder or a flutter on the indicator, but for all practical purposes the telemetering reading is continuous.

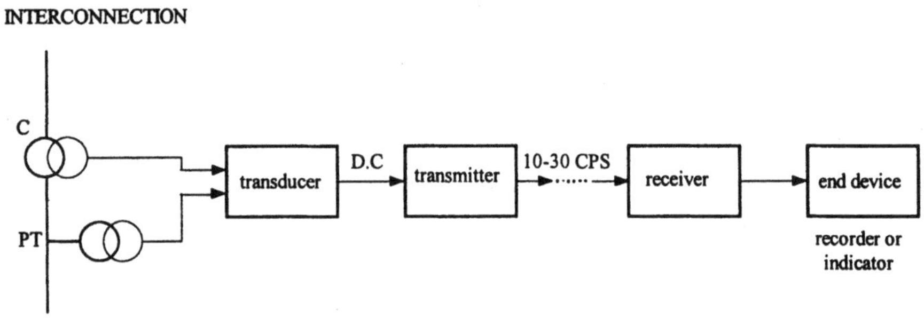

Figure 4.6: *Block diagram of analog telemetering system*

4.5.1.2 Selective analog

When supervisory control is required for remote locations, it is economical to combine an analog telemetering system with control and use it for selective telemetering, as shown in Fig 4.7.

Here two-way communication is involved. Therefore, an analog transmitter can be switched to any transducer by the central operator. This reading is indicated at the central office until removed by the operator or automatically by the supervisory equipment. There is no real limitation to the number of readings which can be selected via supervisory over one channel of analog equipment.

4.5.2 Digital Telemetering

While analog telemetering uses a continuos measure of the concerned quantity, digital telemetering is based on sampling, at fixed time intervals, of the concerned quantity. With the advent of solid state devices, it has become possible to produce digital telemetering systems which have a high degree of reliability. Most of the available pulse-code schemes contain self-checking techniques which will provide an indication at the receiving end if errors have occurred in transmission. Digital telemetering is inherently capable of greater accuracy than analog methods.

With digital telemetering by the pulse code system, the measured value which is first converted into an equivalent direct current is then transformed at the sending end by special analog-digital converter into an impulse code, and is transmitted in a digital form to the Load Dispatch Centre. There, it is either reconverted by a digital-analog converter into an analog value suitable for indication or recording or it is directly fed to an equipment for digital data processing.

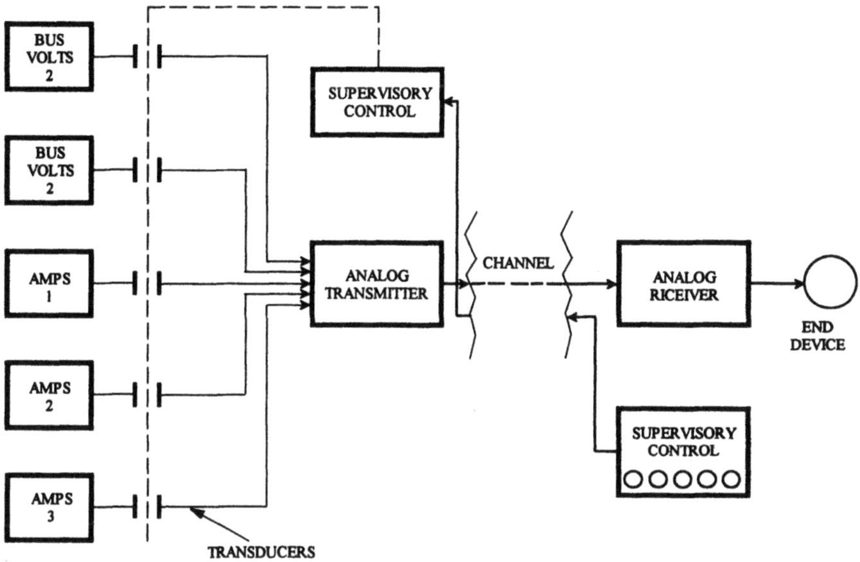

Figure 4.7: *Analog telemetering system with supervisory control*

In the digital telemetering system, the same task of measuring interchanged megawatts is performed as shown in Fig. 4.8. The system starts with the same transducer to convert a.c. MW to a proportional d.c. signal. Now the analog-digital converter takes over, converting the analog signal to various forms of binary codes. Digital transmitter converts this binary readout into a digital code which could be transmitted over the intermediate channel to the digital receiver.

The digital receiver takes the pulse code, verifies it, and converts the message to an output suitable for the end device. If this were a digital indicator, the receiver would present its output in decimal form. For analog indication, a digital-analog converter has to be employed.

4.5.2.1 Digital Cyclic Telemetering

In this form of telemetering, a number of different items of information can be transmitted successively in digital form over telephone wires, carrier links or radio relay links. Transmission may be point-to-point (from A to B), or in a chain (from A to B to C and so on to N).

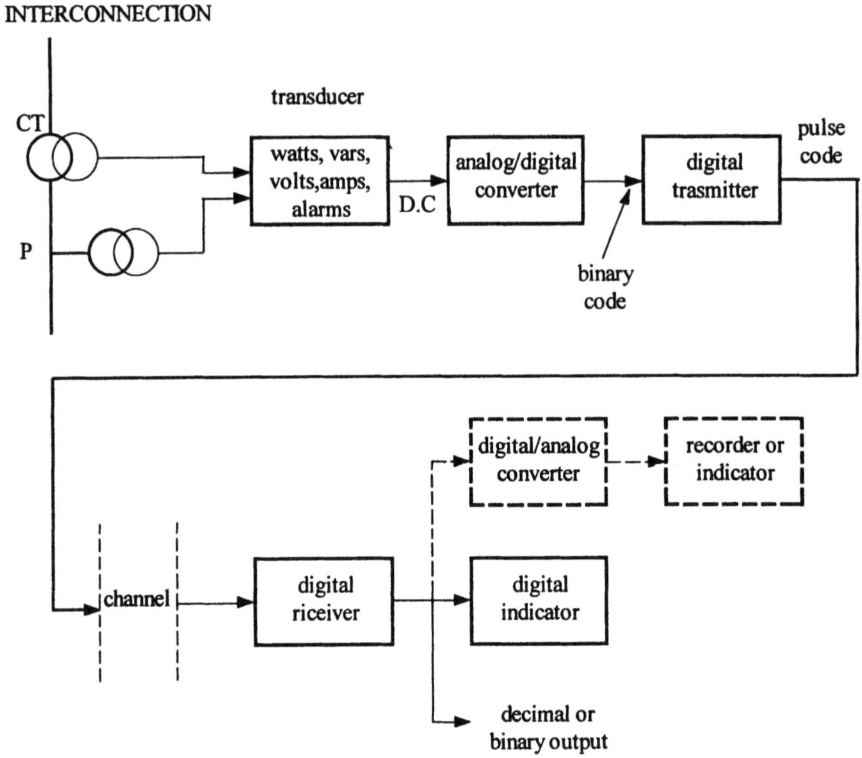

Figure 4.8: *Digital telemetering system*

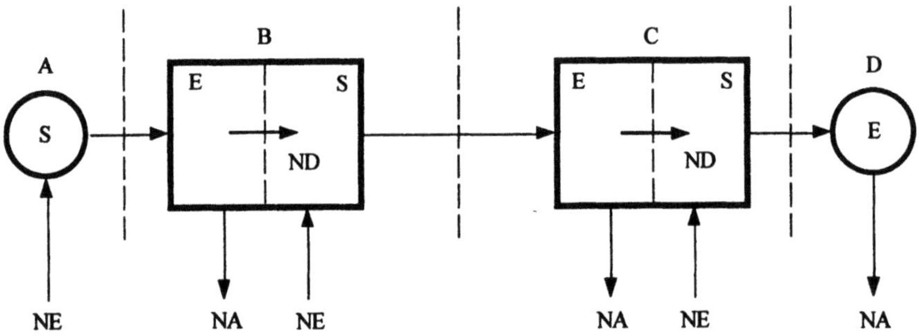

A= Transmitting Station (Information Input Only)

B,C= Transit Station [Input (Ne), Output (Na) And Tr Throughput(Nd) Of Information Possible]

D= Receiving Station (Information Output Only)

S= Transmitter

E= Receiver

Figure 4.9: *Digital Cyclic telemetering system- Chain arrangement*

In the case of point-to-point connections, the information that comes in at A is put out again at B. With a chain, however, information can be introduced at A, or at any of the intermediate stations between A and D and put out again at B or any other transit station, as shown in Figure 4.9. The digital cyclic telemetering system is thus able to gather information along the length of line and to give it up at intermediate stations or at the end of the line.

The digital cyclic system can also be arranged as a radial system, where each of the outstations is able to transmit information to all the others (Fig. 4.10).

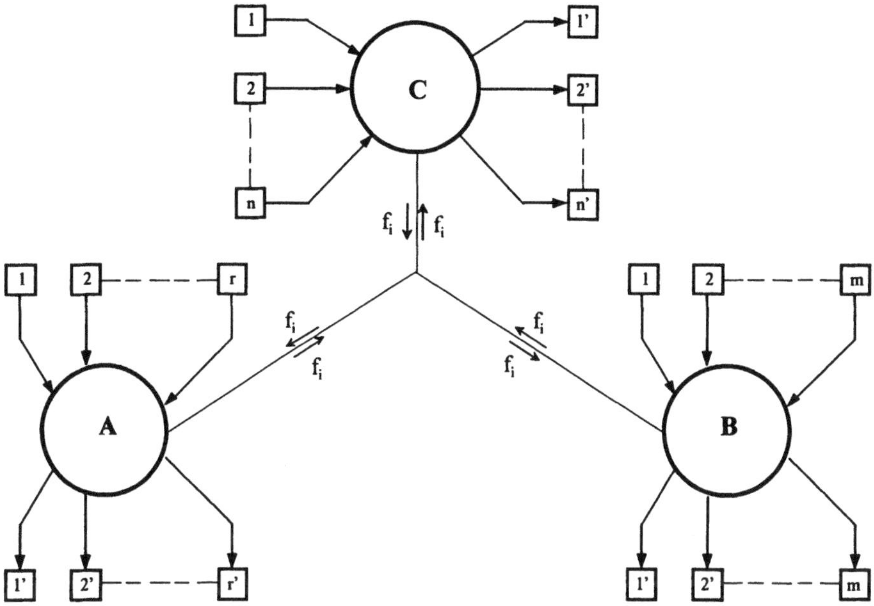

Figure 4.10: *Digital Cyclic telemetering system - Star arrangement*

CHAPTER 5
DETERMINATION OF OPERATING RESERVE

5.1 General

Operating reserve is required to meet the following contingencies:

- rapid fluctuations in demand or outputs of generating sets
- unexpected reduction in the output of generating plant (partial outages)
- sudden loss of generation
- under-estimation of demand or unexpected increase in demand
- unexpected constraints in transmission plant.

Following the loss of a generating unit in one member of an interconnection, governor action throughout the interconnection takes place to restore the frequency (this is commonly called primary regulation); effectively, the deficit of power in the affected system will in the short term be made up largely by transfers from the neighbouring systems. Since the probability of simultaneous loss of generation in all the constituents is extremely small, the ratio of maximum sudden loss of generation to system size is much smaller than for a single system. As a result, frequency deviations for the interconnection are likely to be much smaller than for the single system, but there are likely to be substantial short-term changes in power flows across the interconnection, particularly in the immediate neighbourhood of the member suffering the generation loss.

Under the action of the primary regulation alone all the difference between the actual demand and the generation dispatch will cause a difference in frequency and might result in a risk of overload of the interconnecting links between systems. It is therefore necessary to superimpose on the primary regulation a new regulation, the objective of which is to adjust in each of the individual systems, the power generated relative to the power consumed in order to cancel out the divergencies of exchange power between systems from their programmed values. This is the part played by the secondary regulation, also called tie-line frequency control, which acts in a relatively short time to restore inter-system flows to their target values.

The generating units participating in the secondary regulation do not operate at maximum output. Thus one provides a margin for manoeuvre (the teleregulating range) which operates automatically when needed.

The proportions of hydro-electric and thermal plant in the secondary reserve are governed by economic considerations. They can in fact vary from day to day according to the geographical distribution of the demand. The location of the reserve has also to be checked for a single fault (security n-1) at any point in the system to ensure that it is technically possible to bring all the available reserve to bear.

Faults on the transmission network are generally less serious in regard to the integrity of the total system as in majority of cases they do not affect directly the load-generation balance. The potential risk is that some lines may be overloaded in the depleted network following the fault. In extreme cases this could result in a loss of generation from sequential tripping of transmission lines.

There are three problems which arise in evolving a philosophy on operating reserves. These are the magnitude of the reserve to be held, the time scales of response of components of the reserve and the geographical distribution of reserve throughout the system.

It should also be ensured that the security and quality of supply meet agreed standards in the context of the assumed credible contingencies, that the overall cost of making provision for the reserve and of using it when required is minimised, and that the reserve can be used without risks in the operation of the transmission network.

The general practice regarding the maintenance of operating reserve may be summarised as follows:

Operating reserve consists of Spinning Reserve and Ready Reserve.

The Spinning Reserve component ranges from 50 to 75 per cent depending on the hydro-thermal generation mix during any hour. This is generally equal to, or exceeds, the maximum net demonstrated capability of the largest unit in service in the interconnected system. A good percentage of this - about 70% - must be responsive to load-frequency control without manual intervention. The general practice has been that not more than 15% of spinning reserve is carried on any one machine nor more than 25% in any one power station. Further, spinning reserve carried on one steam unit should not exceed 25% of the net demonstrated capability of that unit.

Ready Reserve should generally equal or exceed the sum of the maximum net demonstrated capabilities of the two largest units in service in the interconnected system. This includes interconnection power immediately and continuously available.

When scheduling the operation of a power system (e.g., daily operation), it is understood that there are uncertainties in the forecasts. In particular, in the load-generation balance which must hold at any instant (generated power equals load plus exported/imported power), all the three components are uncertain.

We shall develop a methodology which allows us to take into account quantitatively their randomness. This will lead to a quantitative evaluation of the reserve necessary to face, with a given risk, the uncertain events ordinarily occurring.

5.2 Availability of Generating units

5.2.1 Availability of a single generating unit

Like every manufactured component, the generating units, hydro or thermal, have a limited life during which they are subject to failures.

During the first period of the life, which is a settlement period, the number of failures per unit time is high; it stabilises next on a lower value (steady state value), and finally it rises again to wear and tear values. We shall consider the central period (steady state), which is by far longer than the other two. Fig. 5.1 represents a plot of the failure rate versus the age of the equipment.

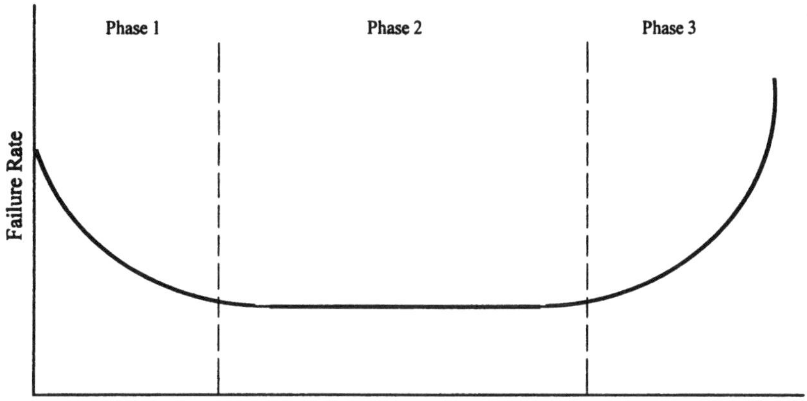

Figure 5.1: *Failure rate versus age of the equipment*

A unit will be characterised by two states: state zero, unit unavailable, and state 1, unit available at full power. A more accurate analysis could envisage, for example, three states: unavailable, available at reduced power, and available at full power, but for the present, we shall limit ourselves to the case of two states.

"Available" does not necessarily mean "in operation"; an available unit can be at rest. If the units have remained as long as possible in state 1, it is necessary that every unit is subjected to a periodical "ordinary maintenance" - the period of overhaul lasting some weeks or a few months - longer for thermal units, shorter for hydro - every year.

The failures which arise while passing from state 1 to state 0, and their type, which determine the period of stay in state 0 for the necessary repairs, can be considered as random events, if there are no design or manufacturing defects, and ordinary maintenance is performed correctly.

It may hence be assumed that :

a) the switching from state 1 to state 0 and vice versa takes place in zero time;

b) the period of stay in each of the two states is finite and non-zero;

c) the average duration (over a long period, for example, one year) in each of the two states is constant over time and is independent of the preceding failures of the unit. The process of passing from one state to the other is "stationary".

The hypothesis c) allows us to define the coefficients a_{01} and a_{10}, constant with respect to time, called "transition rates", in the following manner: if the unit is at time t_0 in state i, $a_{ij}\,\Delta t$ is the probability that the unit switches on to state j during the following infinitesimal time interval Δt; that is the probability that the unit is, at time $t_0 + \Delta t$, in state j.

The probability that the unit remains in state i will then be $1 - a_{ij}\,\Delta t$ ($i,j = 0,1$; $i \neq j$).

Now let $P_i(t)$ be the probability that the unit is in state i at the generic time t ("state probability").

At time $t + \Delta t$ the unit may be in state i as a consequence of one of the following two disjoint events: at time t the unit is in state i (with probability $P_i(t)$) and during the following time

I. during interval Δt, it does not change state (with probability $1 - a_{ij}\,\Delta t$);

II. at time t the unit is in the state j (with probability $P_j(t)$) and during the following time interval Δt, it changes state to i (with probability $a_{ji}\,\Delta t$).

Accordingly, we have

$$P_0(t+\Delta t) = (1 - a_{01}\cdot\Delta t)P_0(t) + a_{10}\cdot\Delta t\cdot P_1(t)$$

$$P_1(t+\Delta t) = (1 - a_{10}\cdot\Delta t)P_1(t) + a_{01}\cdot\Delta t\cdot P_0(t) \qquad (5\text{-}1)$$

As $\Delta t \to 0$, we obtain from (5-1):

$$\frac{dP_0(t)}{dt} = -a_{01}P_0(t) + a_{10}P_1(t)$$

$$\frac{dP_1(t)}{dt} = a_{01}P_0(t) - a_{10}P_1(t) \qquad (5\text{-}2)$$

Since for each t:

$$P_0(t) + P_1(t) = 1 \qquad (5\text{-}3)$$

the integration of equation (5-2) is straightforward (we assume that $P_{00} = P_0(0)$ and $P_{10} = P_1(0) = 1 - P_{00}$ are known):

$$P_0 = \frac{a_{10}}{a_{01} + a_{10}} + \left(P_{00} - \frac{a_{10}}{a_{01} + a_{10}}\right) e^{-(a_{01}+a_{10})t}$$

$$P_1 = \frac{a_{01}}{a_{01} + a_{10}} + \left(P_{10} - \frac{a_{01}}{a_{01} + a_{10}}\right) e^{-(a_{01}+a_{10})} \tag{5-4}$$

Assuming that the unit has switched on to state 1 at time zero, from hypothesis b) listed above, we obtain $P_{00} = 0$ and $P_{10} = 1$ (because the stay in a state has a non-zero duration), and the equations (5-4) become:

$$P_0 = \frac{a_{10}}{a_{01} + a_{10}} - \left(\frac{a_{10}}{a_{01} + a_{10}}\right) e^{-(a_{01}+a_{10})t}$$

$$P_1 = \frac{a_{01}}{a_{01} + a_{10}} + \left(\frac{a_{10}}{a_{01} + a_{10}}\right) e^{-(a_{01}+a_{10})t} \tag{5-5}$$

The equations (5-5) are valid under the hypothesis that the unit has switched back to state 1 at time zero (Fig. 5.2).

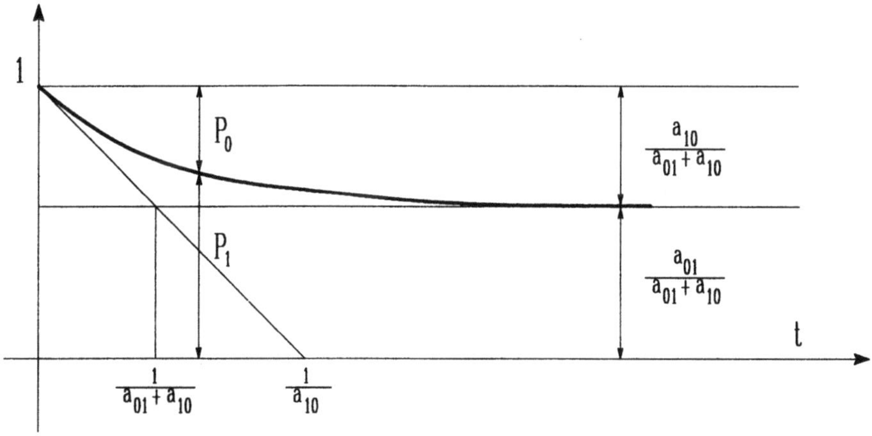

Figure 5.2: *Probability P_1 that a unit be available at time t, supposing that it switched to state 1 (available) at time zero (state probability, medium-long term).*

On the other hand, in a long period of time the probability that the unit, irrespective of its initial state, is in state 1 is the ratio between the total time T_1 elapsed in state 1 during the period recalled above, and the time T_1+T_0, T_0 being the total time elapsed in state 0 (due to failures, that is, excluding the time spent on ordinary maintenance). In a similar manner, the probability of the unit being in state 0 is $T_0/(T_0+T_1)$.

These two probabilities may also be computed from equation (5-4), evaluating the limits for t tending to infinity:

$$\lim_{t\to\infty} P_1 = \frac{a_{01}}{a_{01}+a_{10}} \tag{5-6}$$

$$\lim_{t\to\infty} P_0 = \frac{a_{10}}{a_{01}+a_{10}} \tag{5-7}$$

The first value is called the availability factor and the second the unavailability factor, and hence:

$$\frac{T_1}{T_1+T_0} = \frac{a_{01}}{a_{01}+a_{10}} \tag{5-8}$$

$$\frac{T_0}{T_1+T_0} = \frac{a_{10}}{a_{01}+a_{10}} \tag{5-9}$$

We shall now determine the probability $Q_1(t)$ that the unit, having switched on to state 1 at time zero, remains uninterruptedly in state 1 at least up to time t (that is, without undergoing state changes in the time interval 0 - t). "at least" here means that the unit remains in state 1 for a duration of t or more than t. Obviously:

$$Q_1(t+dt) = Q_1(t)(1-a_{10}t),$$

from which follows:

$$Q_1(t) = e^{-a_{10}t} \tag{5-10}$$

Note that this probability , different from state probability P_1, tends to zero for t tending to infinity implying that the unit will certainly undergo failure as t tends to infinity (Fig. 5.3).

The probability that the unit will be in operation without interruption for a time period between t and t+dt is $Q_1(t)$-$Q_1(t+dt)$ which by definition is $-Q_1'dt$ (Q_1' being the derivative of Q_1). From (5-10) this probability is $a_{10}e^{-a10t}\,dt$.

The average period of uninterrupted operation can therefore be given by :

$$\int_0^\infty t \cdot a_{10} e^{-a_{10}t} dt$$

Performing the integration we obtain $1/a_{10}$.

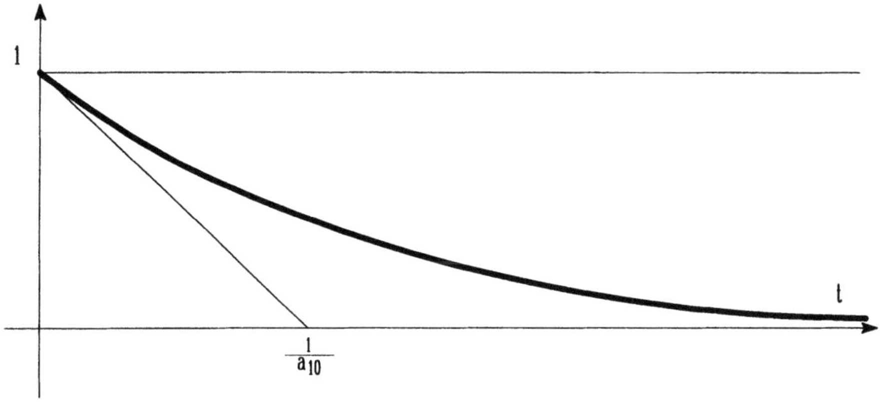

Figure 5.3: *Probability of uninterrupted permanence in state 1, Q_1, assuming that the unit is switched to state 1 at time zero (short term probability).*

On the other hand, the average time of uninterrupted operation is of course equal to T_1 / n, n being the number of state changes from 1 to 0 (and vice versa) during $T_1 + T_0$.

From (5-8) and the above considerations, we obtain,

$$\frac{a_{01}}{a_{01} + a_{10}} = \frac{T_1}{T_1 + T_0}$$

$$\frac{1}{a_{10}} = \frac{T_1}{n},$$

from which we have:

$$a_{10} = \frac{n}{T_1}$$

$$a_{01} = \frac{n}{T_0} \qquad (5\text{-}11)$$

The values of T_1, T_0 and n are obtained experimentally. T_1 / n is called "mean time between failures" (MTBF) and T_0/n "mean time to repair" (MTTR).

Practical values of the availability factor defined above are 0.9 - 0.95 (higher values for hydro units, lower for thermal). Typical MTTR values are 24 - 120 hours for hydro units, 72 - 480 hours for thermal.

Hence a unit having an availability factor of 0.95 and an MTTR of 120 h has transition rates $a_{01} = 1/120\ \text{h} \cong 0.00833/\text{h}$ and $a_{10} = (0.05/0.95)/120\ \text{h} \cong 0.000439/\text{h}$. In a period of 100,000 h (excluding the time devoted to maintenance), it is available for 95,000 h and unavailable for 5,000 h (due to failures). It undergoes 0.00833 x 5,000 $\cong$ 41.7 switchings from state 0 (unavailable) to state 1 (available) and 0.000439 x 95,000 $\cong$ 41.7 switchings from state 1 to state 0 (of course in a number equal to that of inverse switchings).

Again, with availability 0.95 but MTTR 48h we obtain $a_{01} \cong 0.020833$/h and $a_{10} \cong 0.001006$/h, with 104.1 switchings in one sense and in the opposite, in 100,000 h.

With an availability of 0.98 (very good) and MTTR 48h we have $a_{01} \cong 0.020833$/h and $a_{10} \cong 0.000425$/h, with 41.7 switchings in each direction in 100,000 h.

5.2.2 Availability of a set of generating units

The expression (5-10) of Q_1 is not of much practical utility to evaluate the availability of each single generating unit for the following day, because we should evaluate the occurrence of a single event and the reliability of such a forecast would be very low.

On the contrary, (5-10) is useful to evaluate the availability of the whole set of units, following the procedure described here. We shall refer to probabilities Q_i and not P_i because for a unit becoming unavailable at the time of preparing the forecast for tomorrow, it is impossible, due to the order of magnitude of MTTR mentioned above, to be again available tomorrow.

While preparing the forecast for the following day, we know how many and which units would be in operation at the peak hour of the following day itself. It is also known from how much time each of them is available (that is, from how much time each of them has undergone the last switching from state 0 to state 1).

Let m be the number of interested units, and t_{0i} (i = 1,...,m) the time from last switching to state 1 to the peak hour of tomorrow of the i-th unit.

For each unit, the value $Q_{1i}(t_{0i}) = Q_{1i0}$, that is, the probability of being again available at the peak hour of tomorrow, can easily be computed from (5-10).

The decision as to whether the generic unit is available or not is left to chance, since the occurrence of a failure is a random event ; in any case, it will be necessary to consider that the probability of availability is higher for the units having a higher Q_{1i0}.

One can then, for each of the m units concerned, draw in a random fashion and with uniform probability a number between 0 and 1; the unit will be considered available if the number x_1 is lower than Q_{1i0}, unavailable if it is higher.

The set of m drawals gives one a particular value of unavailable power Pj (the index j will be explained later), MW, which is the sum of the rated powers of the units drawn as unavailable (that is, of the units for which x_i turned out to be higher than Q_{1i0}).

One then repeats M times the set of m drawals, and a set of M values of unavailable power Pj (j = 1,...,M) is built up. By arranging the M values of Pj in increasing order one obtains a curve like the one shown in Fig. 5.4 in which the ordinates are the percentage number of drawals (over M) and the abscissae the value of Pj which correspondingly (that is, for that percentage of times) has not been exceeded.

In other words, from Fig. 5.4, it turns out that in 100% of the M drawals the unavailable power has been less than or equal to pj100; in 90% of the drawals it was less than or equal to pj90, and so on.

The curve of Fig. 5.4 is interpreted as a probability curve: for each value of unavailable power in the abscissa, the corresponding ordinate value represents the probability that it is not exceeded; that is, for example, pj90 has a probability of 90% of not being exceeded.

It is clear that, disregarding other sources of uncertainty, to be certain that every chance event concerning the availability of generating units will be covered, we should assume a reserve equal to pj100; if, on the contrary, we accept a risk of 10%, we shall assume pj90, and so on.

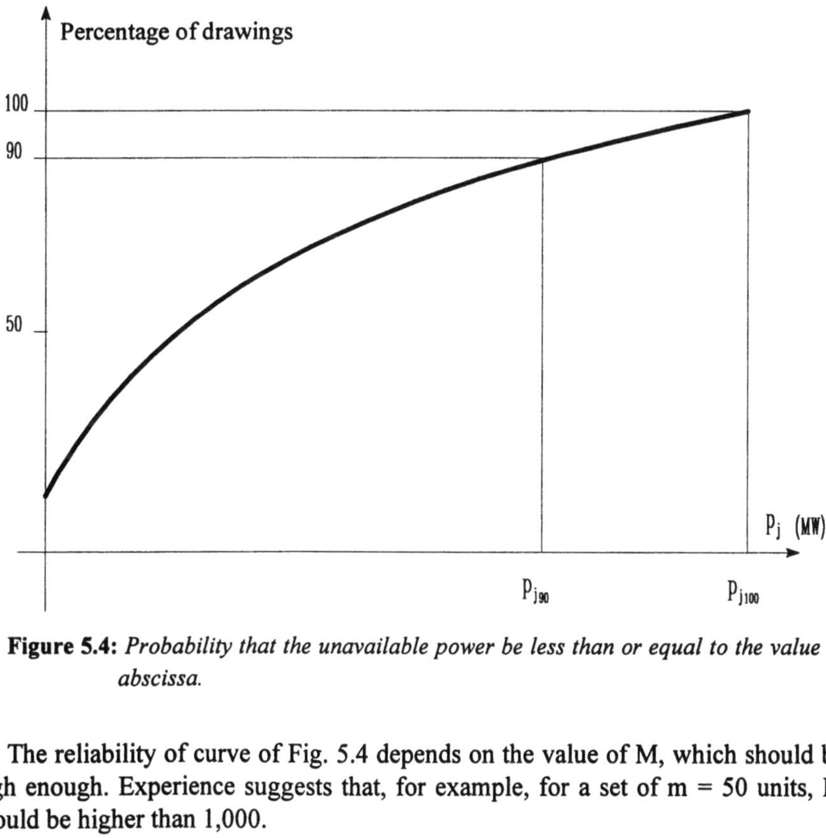

Figure 5.4: *Probability that the unavailable power be less than or equal to the value in abscissa.*

The reliability of curve of Fig. 5.4 depends on the value of M, which should be high enough. Experience suggests that, for example, for a set of m = 50 units, M should be higher than 1,000.

5.3 Load uncertainty distribution

5.3.1 Load forecast errors

We shall focus our attention on the forecast of the peak load; the method developed for the peak load can be repeated for the load at any other hour of the day.

We recall that the forecast concerns a "smoothed" value of the load, in the sense that the actual load has rapid fluctuations around a mean value holding, for example, for 5 or 15 min; this mean value is the value actually forecast (the day before), and recorded (after the fact).

We shall suppose that historical series of forecasted and actual load (MW) are available in the records; hence one can easily compute their difference (forecast minus actual), constituting the historical series of mean load forecast errors e_{LF}.

The values of e_{LF} may be grouped in classes (e.g., from 1 to 10 MW), and a frequency curve like that shown in Fig. 5.5 is obtained; this curve is interpreted as a probability curve, that is, it is assumed that in the future the forecast errors will behave according to that curve.

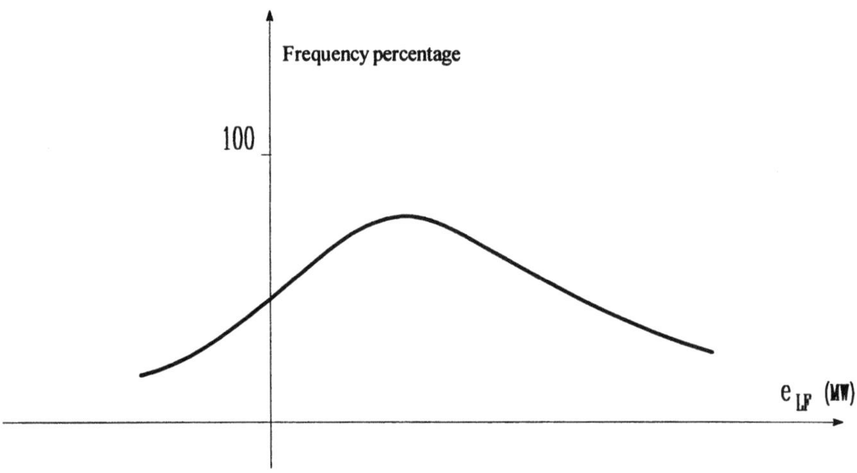

Figure 5.5: *Frequency distribution of load forecast errors*

5.3.2 Random fluctuations of load

To recall from the preceding section, the load has random fluctuations around its 5 to 15 min. mean value; they depend on the fact that some elementary electrical loads of the network connect and disconnect themselves in a random manner. There is an empirical expression for the amplitude p_{RF} of this fluctuation, given by

$$p_{RF} = k_{RF}\sqrt{P} \tag{5-12}$$

where p_{RF} and P are in MW, P is the load, and k_{RF} = 1 - 2 (heuristic value).

The fluctuation is less than proportional to the load, because as the size of the network increases, the compensation effects also increase.

The quantity p_{RF} may be considered to include also the random variations of exchanged power in the case of interconnected system.

5.4 Determination of necessary reserve

Monte Carlo method, similar to the one used to build up the curve of Fig. 5.4 will be followed:

1) a set of m drawals (m = number of generating units) is performed, like that described in section 5.2.2, and the value P_A of available power is computed, equal to the difference between the sum of the rated powers of the units being not under maintenance and the sum of the rated powers of the units which turned out to be unavailable in the set of m drawals;
2) a drawal is performed, with a probability distribution equal to that determined in section 5.3.1 (Fig. 5.5) of the load forecast error e_{LF};
3) the computation is performed of the deficit $D = L_p - e_{LF} - P_A - P_i$, where L_p is the peak load forecast (known constant quantity) and P_i is the imported power (known constant).

The three operations 1), 2) and 3) are repeated an appropriate L number of times, obtaining L values of D; these values are sorted out in increasing order and cumulated; a curve like that of Fig. 5.6 is built up, which gives the probability of not exceeding the generic value of deficit, taking into account the randomness of both availability of generating units and peak load forecast.

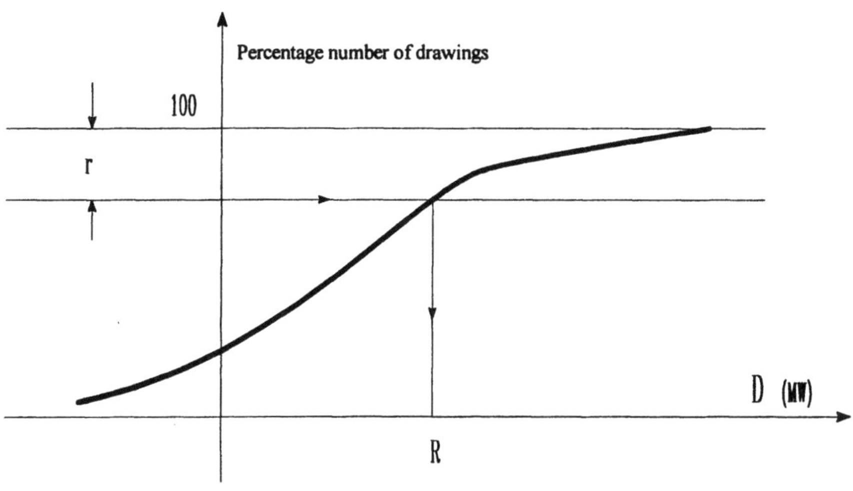

Figure 5.6: *Probability of not exceeding the deficit value in abscissa.*

In other words, the curve of Fig. 5.6 can be interpreted as follows: the deficit R (MW) has a probability (100 - r)% of not being exceeded, that is, in (100 - r)% of the possible cases, the power deficit will not be greater than R.

If r% is considered as an acceptable risk, then R will be the reserve to be made available. It will be necessary to add another appropriate quantity to this last value (for example, just p_{RF}) to account for the random fluctuations of the load (section 5.3.2); let S (MW) be the final value of reserve so determined.

This procedure may be repeated, for example, for the hour of minimum load of the day, determining another value of reserve which probably will be less than that valid for peak hour; these two values may be assumed valid, one (that of peak hour) for the hours of "full load" (normally, for example, from 8 a.m. to 6 p.m.), the other for the remaining hours of the day.

5.5 Classification and assessment of reserve

In actual practice, the determination of the amount of necessary reserve S is often much simpler than that illustrated in section 5.4. For example, provision is made for a reserve equal to the rated power of the largest generating unit in parallel, or to a percentage (2 - 6%) of the peak load, or the largest of the two increased by a suitable quantity taking into account the random fluctuations of the load, etc.

Not all of these simplifying assumptions have a sound theoretical or practical background; in particular, the assumption of a certain percentage of load seems somewhat arbitrary. The reserve, determined with the methodology illustrated in section 5.4, is called "operating reserve" S required for the daily operation; there is also the so-called "medium-term" reserve, required to cover a longer term and is determined by taking into account other factors (such as maintenance requirements, etc.).

The daily operating reserve is that amount of generating capability in excess of a system's internal peak load minus interruptible loads, which is available to provide for load variation and forecast error, frequency regulation, and contingencies such as loss of generating capability. Here, internal load is the summation of the net output of the system's generators plus the net of interconnection receipts and deliveries; and interruptible load is a consumer load contractually subject to interruptions when required and pumped storage hydroelectric generation in the pumping mode. Further discussion on interruptible load is covered in the next section.

The operating reserve must be developed at the most in a few tens of minutes; and the medium-term reserve in some hours (e.g., cold thermal units), which is also called "cold slow start reserve".

The operating reserve is in turn divided into "spinning reserve" developable in a few minutes - generally within 10 minutes - spinning thermal and hydro units, and "standby quick start reserve", developable in a time ranging from a few minutes to a few tens of minutes (for example, Diesel, hydro and turbogas). The standby quick start reserve is also called "minute reserve" which must face the internal disturbances after a few minutes of help from neighbours ("equity" principle).

Hence, not all the operating reserve S is spinning, either because the part depending on load forecast error requires slow intervention, or because if the network is interconnected, it receives a quick initial assistance from the other partners, which may be compensated relatively slowly afterwards ("solidarity" principle). For an isolated (not interconnected) network, on the contrary, it will be wise to have all the operating reserve S as spinning reserve.

When a disturbance occurs (for example, the tripping of a generating unit) and the spinning reserve is consequently engaged, the spinning reserve itself should be built up again by starting up the units constituting the minute reserve.

The spinning reserve is again divided into "primary reserve" (developable in a few tens of seconds) and "secondary reserve" (developable in a few minutes). The primary reserve is that part of the spinning reserve which is controlled by primary frequency regulators, and the secondary one by secondary load-frequency control.

For example, in a spinning reserve of 100 MW, say, 10 MW is subjected to primary control and 90 MW to secondary control.

Spinning reserve and minute reserve are approximately of the same quantum.

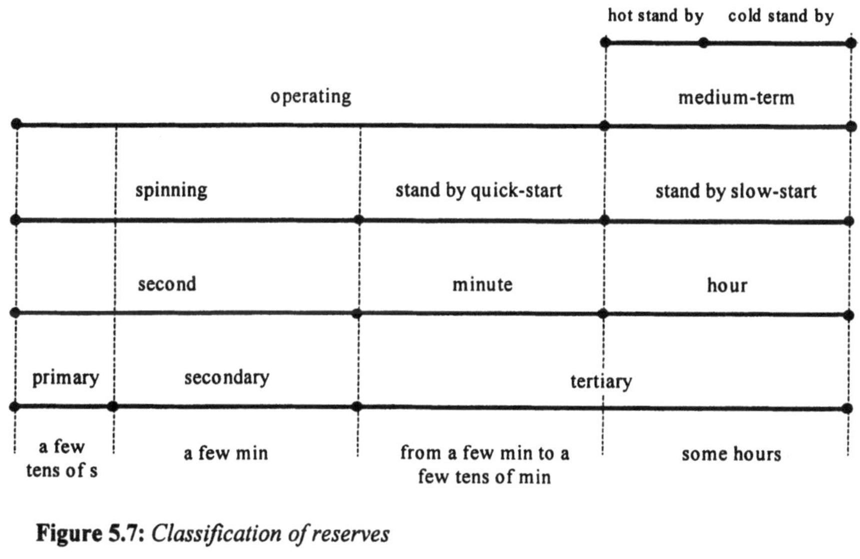

Figure 5.7: *Classification of reserves*

There is also a so-called "tertiary reserve", which is the sum of the standby quick or minute reserve plus standby slow start or medium term reserve. Some utilities further subdivide the medium-term or standby slow start or hour reserve into "hot standby" (conventional thermal units in standby but with burners of the boilers at minimum firing, in order to keep the unit ready to start) and "cold standby". The hot standby reserve is also called "contingency reserve" and should be used to meet the uncertainties in planned unavailability (maintenance) of generating plants.

For considering the availability of reserve, a given available unit should be able to develop its rated power for a duration of at least, say, 4 hours (this value varies from utility to utility); for thermal units this is supposed to be always fulfilled (a sufficient availability of fuel is assumed); for hydro units the feeding pond or reservoir must have a sufficient quantity of water available.

Fig. 5.7 illustrates the various definitions and the characteristic time durations given in this section.

5.6 Assignment of reserve to various units and resources

In general, the reserve should be distributed over as many units as possible. This is particularly true for spinning reserve in order that it may develop faster; because of technical reasons, the contribution of the units to spinning reserve is restricted to a band of power lower than the rated power of the unit.

It may be recalled that a conventional thermal unit varies its output at a maximum speed of the order of 1%/min of the rated power, from the technical minimum to the rated power; on the other hand, in a restricted band of variation (for example, between 90 and 100% of rated power), the speed of variation (up and down) may be much higher (e.g., ten times more). In such a limited band nuclear units can also satisfactorily contribute to spinning reserve.

A hydro unit has a maximum speed of variation of the order of 1%/s on MW output, from technical minimum (of the order of 30% for Francis type turbines, and zero for Pelton) to rated power.

For the same technical reasons which limit the maximum speeds of variation of power output, when assigning a thermal unit to spinning reserve duty, its regulation band is fixed at a value not higher than about 10% of the rated power; for the hydro units, the regulation band may be higher -- from 30 to 100% of rated power or more, according to the type of turbine.

There is also a source of reserve available to the operator in load management. This consists in the possibility for the operator to shed some load temporarily as a normal operation action. The reserve implemented through load management may be considered as a primary reserve. However, this practice is not universal at present, but there is an increasing trend to adopt this technique.

The pumped storage plants, when pumping, may be considered as a primary reserve as they can be shed in a very short period of time.

These two forms of primary reserve are also called interruptible load.

In the case of interconnected systems, there may be an agreement between the utilities for sharing the operating reserve: Utility A, for which the computed operating reserve is S (MW),has $T<S$ reserve located in its own network and $S-T$ available on request through tie-lines in another interconnected system (utility B).

In the case of hydro-thermal systems, the question may arise as to which units, thermal or hydro, should be entrusted the spinning reserve task, from an economical point of view. If the regulating bands are those mentioned above --10% for thermal units and 70 to 100% for hydro, it is more convenient to run hydro units at the scheduled values and entrust the regulation task to thermal units. The reasons are: the efficiency of hydro units is low at low loads and the thermal units have good efficiencies. Thus the problem of allocation of spinning reserve to various units requires a detailed examination in each particular case.

5.7 UCPTE philosophy

The basic philosophy in the UCPTE countries for provision of reserves is as follows:

First level

Primary regulation takes place through governor action on turbines, actuated by a frequency deviation from the set point. This could be supplemented by change in the mode of operation of pumped storage stations. It is obvious that only units which are already in synchronism with the system can provide this reserve. For example, the machines (300 MW each) at Dinorwic pumped storage station in the U.K. (1800 MW) are capable of being loaded from synchronisation to maximum output in about 10 seconds (incidentally, the U.K. does not belong to the UCPTE).

Second level

- load-frequency regulation (adjustment of the generation of a number of steam sets or hydro-electric sets to satisfy the set values of frequency and of inter-system exchanges)
- start-up of gas turbines by under-frequency relays.

Third level

- adjustment of generation of quick-start hydro-electric stations or pumped storage stations
- start-up of gas turbines
- adjustment of the generation of steam sets not operating at maximum output
- synchronising sets held in a "hot" state
- emergency exchanges with other countries.

The overall capacity of the Continental European system is sufficiently large for the frequency drop to be always very slight. In fact, this frequency drop Δf is linked to the loss of capacity ΔP by the formula $\Delta f/f = \sigma \Delta P/P$ where σ is the power-frequency characteristic or permanent droop of the system (on an average about 13 per cent). For example, at the peak load of about 250,000 MW on the UCPTE system (in 1994), the loss of 1000 MW results in a frequency drop equal to $\Delta f = 0.13 \times 1/250 \times 50 = 0.026$ Hz.

Sharing of reserves in the UCPTE countries

The following arrangements for reserve sharing between the different utilities in the UCPTE interconnected system are in vogue:

- $t_1 \leq 0.5$ -10 min : <u>Momentary reserve</u>

  For the first few minutes after an outage, the power deficit is covered by the momentary reserve of all partners in the interconnected system (solidarity principle). This is provided by the primary governor action of all operating units.

- 0.5 - 10 min < t_2 ≤ 0.5 - 10 h : Fast reserve

  Within a short time, ten minutes at the most, the system causing the power deficit is expected to cover it from its own reserves (secondary control action), thus relieving the supporting partner systems (equity principle). For this purpose, fast reserve with a mean access time of 0.5 - 10 minutes must be available. It normally consists of spinning reserve, gas turbines, pumped storage plants and hydro units.

- 0.5 - 10 h < t_3 mean repair time : Slow reserve

  The fast reserve must, however, be replaced as rapidly as possible by slow reserve (cold thermal units and large long-term storage) because of its higher operating costs and limited capability, and because of the need for the fast reserve to be available in case of further outages. The slow reserve has a mean access-time of 0.5 - 10 hours.

The values of reserve expressed in percentage of the peak load in the ENEL system (Italy), a constituent of the UCPTE, are as follows:

The total operating reserve (cold reserve not included) represents about 5 to 6 per cent of the peak, of which:

- about 2.5 to 3 per cent is "second" reserve, in any case not less than the power of the largest running unit,
- the remaining 3 per cent is "minute" reserve, in any case not lower than the "second" reserve, for replacing it if necessary,
- 1.5 per cent of reserve is under LFC; this means that the range available is of ± 1.5 per cent. This value is chosen from experience,
- the allocation of the reserve is 50/50 respectively on the hydro and on the thermal plants with some variation during the seasons; 60 per cent and 40 per cent is the subdivision between the Northern and the Southern areas respectively.

The trend is towards increasing participation of thermal plants in the daily scheduling of operating reserve. This is to allow the greatest number of hydro plants to run in accordance with their forecast modulated schedules.

Evaluation of operating reserve and its costs is done a number of times in a day with a view to reducing the uncertainty in the forecast of operating reserve and making an economic analysis of the operating reserve.

References

1. K.W. Edwin and H.D. Kochs of CIGRE Study Committee No. 32, Electra No. 76.

CHAPTER 6
LOAD-GENERATION BALANCE

6.1 Introduction

This chapter is intended to illustrate the methodologies adopted by the power supply industry to set up generation strategies or schedules, taking due account of the requirements for economy and security of operation.

The variety of methods used depends on the mix of the generating systems (hydro and thermal), the importance of the energy exchanges with interconnected utilities, and the presence of opportunity or "secondary" loads or markets.

An important aspect in the algorithmic approach is the choice of simplifications to be adopted in order to reduce memory and computer time requirements, or formulate a mathematical algorithm which could solve the problem even by iteration.

We will deal only with some methods which are more widely used and which have proved effective over time.

The time span considered is from 1 to 10 days ahead for short term scheduling and from 1 month to 5 years ahead for medium-long term operations planning.

Particular attention is paid to the methods adopted to account for randomness, with reference to the particular time span considered and mix of generating systems.

To manage the operation of their power systems most utilities prepare long- and short-term plans with interaction between the two, to meet forecast demands according to the required economic and security criteria. The long-term plans may include many planned activities such as outages of generation and transmission, fuel procurement, coast down of nuclear plants, operation of hydro valleys and storage, and trading arrangements together with the management of plant capacity and various budgetary provisions.

To assess the utilisation of the power plants, some kind of long-term scheduling is usually carried out taking account of these plans to determine system operating and marginal costs, water values and merit orders for various system operating strategies. These in turn are used to tune long-term plans and provide the basis for short-term programming plans.

As the short-term approaches it is necessary to provide the staff concerned with system control, power stations and transmission with guidance of system requirements to meet expected demands and to implement the long-term plans and associated modifications. The problem facing the utilities is how best to manage this and the general approach adopted is to partition the planning timescale, prepare distinct scheduling plans plans for one day, one week or one month ahead with longer term plans covering the year ahead and beyond. The interaction between the two sets of plans may take place by reviewing the long-term plans in the light of the actual operating experience as the need arises or at predetermined time intervals.

There are many methods for preparing short-term generation schedules. These range from scheduling plants based on a simple merit order ranking to meet selected demand peaks and troughs or demand duration curves, to sophisticated computing techniques predicting the generation requirements for each hour of the scheduling period based on both forecasted data and actual operating history.

Some of the features which may need to be considered when developing a new scheduling procedure include consideration of:

- the main characteristics of the power system: thermal, hydro or a mixed system; the system control organisational aspects; data input and output for national and a number of regional control centres;
- the problems unique to the system: system size, demand shape, number of power stations and generators, availability of fuel and its transport, hydro inflows, storage ponds and reservoirs, trading arrangements, transmission constraints and cycling of nuclear plants;
- the length of the scheduling period and elementary time step; generating plant characteristics, ramp up and down times, start up costs, the need for incremental and decremental cost information; output summaries, the practicality of the results, whether the program will be used to update schedules on the day in control room, management and update of data input;
- the need to cover random events, modelling of operating reserves, using a deterministic or stochastic technique, and the skill of personnel to use the program. Some of these issues are covered in Appendix - 2.

6.2 Short-term scheduling

The short-term scheduling period is generally one week; often the schedules, which are determined, e.g. on Friday, cover up to 10 days ahead, to better account for flexibility of thermal units and for particular scheduling needs of week-ends (including occasional maintenance works).

In some cases the short-term schedules are updated during the week, both at the day ahead stage, and a number of times on the day in the control room [1].Hence it is important to have a carefully thought-out procedure which may readily be rerun by control room staff at predetermined intervals or when required. Due to the restricted span of the time considered, the forecasted values (load, water inflows, availability of generating units) are considered deterministic: some allowance for randomness is accounted for by considering the spinning reserve often equal to the largest generating unit in operation.

6.2.1 Thermal Power Systems

We call thermal power systems those systems with no medium or long-term hydro storage (but having perhaps some short-term pond storages or pumped plants).

It may be noted (App. 2A, sec. 5.1) that the input to short-term scheduling in this case comprises weekly totals of thermal generation, opportunity import and export, and secondary load; these totals are in some cases previously partitioned in daily totals (or daily totals are directly produced by medium-term scheduling procedure, for the next week).

The only integral type condition present in this case is a soft constraint (see eq. 8 in App.2A).

Here the short-term scheduling of thermal power systems turns out to be almost independent of medium-long term operations planning, at least in respect of energy allocation. The close connection between short- and medium-term is constituted by the merit order listing of generation and planned non-availability schedule of generating units (trading agreements and forecasted changes in prices and availability of fuel are also important).

Thermal plant merit orders may be derived by ranking all the units in a cost order based on heat cost times heat rate. Two types of merit order are used based on:

1) full load heat rate times heat cost to schedule plant on or off ("specific cost");
2) incremental heat rate times heat cost, to allocate load to plants already operating in the system ("incremental cost").

The merit order cost for scheduling plant to be synchronised to the system may be adjusted to reflect the start-up cost of the marginal units being shared over the amount of energy produced by each unit for a given period of time. In a similar manner the merit order cost of plants to be shut down over night may be discounted by the cost of starting up again. To ensure that the cost of starting up unit again in the morning does not exceed the benefit of shutting down over night, the merit order cost in (1) may be adjusted to reflect the cost of starting up again; e.g. for a 500 MW generating unit costing some L 5000 to start up after a 5 hours shutdown, a discounting may be calculated as,

$$\frac{5000}{5 \times 500} = 2.0 \ \frac{\mathrm{L}}{\mathrm{MWh}}$$

and this is deducted from type (1) costs. This concept may be extended to reflect other penalty costs associated with plant flexibility parameters which do not meet the ideal system requirements, e.g. for peaking plant with long minimum run times of up to some 8 hours compared with 2 hours ideal on-times. Penalties for the out-of-merit period when the plant is not required (i.e. 8 - 2 = 6 hrs) can be assessed and plant costs adjusted as required.

The heat costs are usually derived at the long-term scheduling stage taking account of planned generator overhaul programmes, fuel supply agreements and trading arrangements to meet the expected demand. A number of methods may be used for assessing heat costs for merit orders. These include using the average cost of all fuels assigned to each station, or the marginal fuel that would be assigned to each station to meet the increase in system load forecast. Thus an important link with long term plan is established. In the short term it is necessary to carefully consider the use of any significant amount of cheap spot fuel purchase. This is because penalties may be incurred on longer term fuel contracts, if these are broken or renegotiated.

As far as nuclear generation is concerned, it may be noted as an example that a nuclear PWR (pressurised water reactor) unit is a stock of limited energy. Indeed, unlike fossil fuel powered units, each reactor must be shut down during approximately two months for refuelling, between the given limits of burn-up of nuclear fuel.

As a consequence, from the point of view of operational planning, a PWR unit is similar to a large hydro reservoir without inflows: future values of the energy stored are computed for each unit on a medium-term basis. They are then used as a merit order in the short-term scheduling to help decide nuclear power modulation.

6.2.1.1 Unit commitment

The part of the load diagram to be covered by thermal generation is supposed to be known (average MW in each elementary time interval or step, ETI). For more clarity it is appropriate to distinguish, within the unit commitment algorithm, two different aspects which are, for each ETI:

- the choice of units to be synchronised;
- the share of MW between the running units.

The first aspect involves only the units which may or may not be started up or shut down, and hence does not involve the "shall run" units. (See App. 2 sec. E for the definition of "shall run" units); the second one involves all the running units (including "shall run").

The most important characteristics to take into account for each unit (excluding "shall run" units) are:

- minimum on and off times;
    - start up costs (function of down time duration);
    - and for all units (including "shall run"):
- ramp duration times;
- input/output curves (MJ/h versus output MW);
- available capacity (MW), min and max.

It is worth nothing that normally the number of "shall run" units is high: in addition to the units which must run due to other than economic reasons (if any), the most economic units are likely to be kept on line all through the week (including nights and week end).

This is especially true if the age of the thermal units is staged in a more or less uniform way, as it is natural in almost all power systems (newer units tend to be more economic).

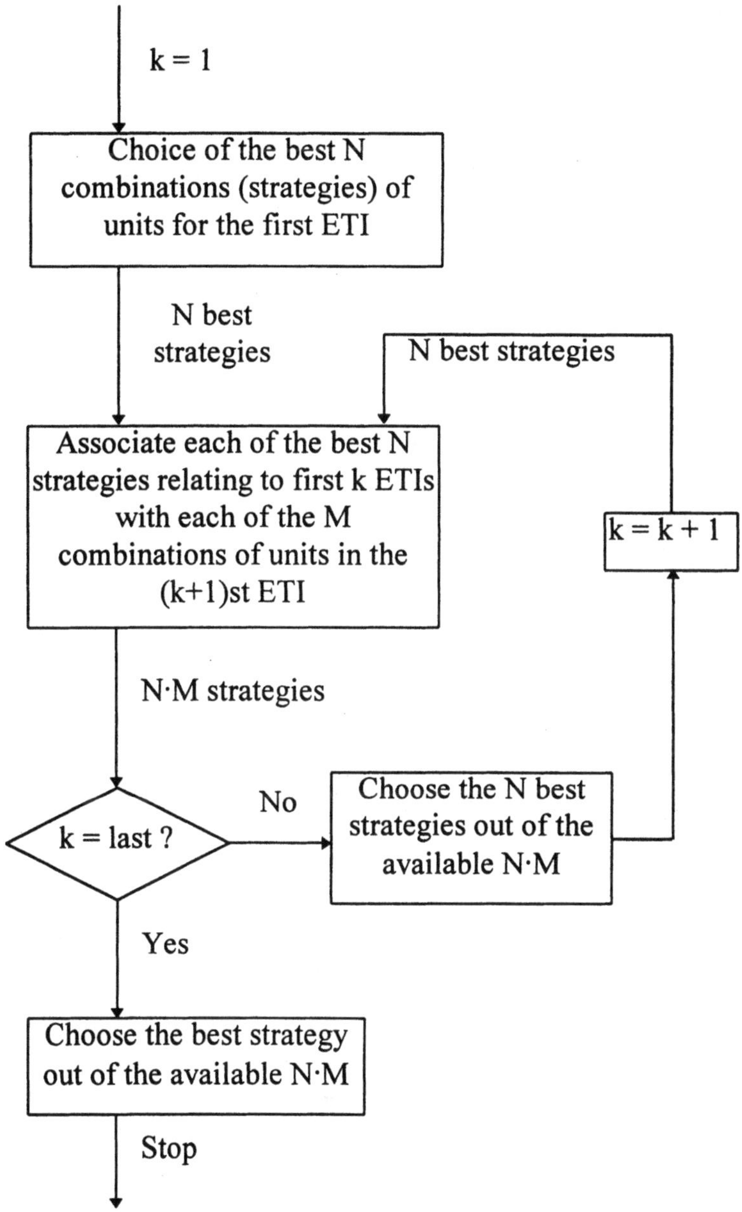

Figure 6.1: *A unit commitment procedure, based on forward dynamic programming*

For example, it may well be that out of 100 available thermal units, 70 may be considered as running anyhow ("shall run").Hence the unit commitment problem involves only a part of the available units.

This is of particular importance, because the unit commitment problem is one which is typically conditioned by dimensionality. Power flow constraints on transmission system may be represented by dividing the system into zones, the inter-zonal power flows being limited at predetermined levels consistent with operational security standards.

An "exact" algorithm would require a combinatorial approach, which is normally impracticable; approximate methods are normally used instead.

Fig. 6.1 shows an algorithm for unit commitment which uses forward dynamic programming; for simplicity, partitioning of transmission network is not considered.

The total number of thermal units is n_t; the number of "shall run" is n_t - m_t; the "free" units are m_t.

Suppose that, for the first k ETI (elementary time intervals) of the period under consideration (e.g. one week), the N lowest cost strategies have been determined. Strategy means here the set of output values (MW) of the thermal units, in each one of the first k ETIs. Some output may of course be zero (unit not committed); the generic strategy is identified by $k*n_t$ output values.

To find out N lowest cost strategies with k = 1 (that is in the first ETI) is a trivial task.

The problem is now to determine a set of N lowest cost strategies for the first k+1 ETIs.

This will be done by the following steps:

1) consider one generic strategy out of the N determined for the first k ETIs;
2) select for the $(k+1)^{th}$ ETI a number of combinations of the m_t free units, complying with the constraints (besides others, that of covering at least the load plus the requested spinning reserve). Let M be the number of permissible (and selected) combinations;
3) for each one of the M strategies of the $(k+1)^{th}$ ETI determine the optimum sharing of MW between the n_t - m_t "shall run" plus the chosen units (e.g. with the criterion of equal incremental costs, with transmission losses accounted for by means, for example, of B coefficients). Determine also the corresponding cost (this will include the start up cost of those units, if any, which are at standstill in the kth ETI, and running in the $(k+1)^{th}$);
4) add up this cost to the cost of the generic strategy referring to the first k ETIs, selected in step 1). M strategies, and corresponding M costs, are so determined for the first k + 1 ETIs;
5) then repeat the steps 1), 2), 3), 4) for all the other strategies referring to the first k ETIs; at the end N*M strategies, with corresponding costs, are available for the period covered by the first k + 1 ETIs;

choose, out of the above N*M strategies, the N with lowest cost ones. Then go to 1), considering the next k.

When considering the last ETI only one strategy (that of minimum cost) will be retained, and this will be the schedule of thermal units.

Note that when selecting units for the $(k+1)^{th}$ ETI (step 2), one of the constraints to consider is that of ramp duration. If the unit was out of service in the kth ETI, it may happen that it will be limited, when sharing power between the units (step 3), to a certain maximum output in the $(k+1)^{th}$ ETI. This is due to its limited ramping speed. Again, a given unit shall not be selected in the $(k+1)^{th}$ ETI if it was out of service at the kth, and its minimum down time has not yet elapsed; etc.

It is easily understood that the procedure described is good enough, provided that N (the number of strategies retained each time) is large enough. "Large enough" is a matter of experience with the various mix of thermal units set. For example, if there are units with long ramp duration and high start up cost (but with low specific cost in normal operation), with a too low value of N they may not be included in the final schedule, even if their operation would have been economically convenient.

As far as opportunity purchases are concerned, these may be represented, ETI by ETI, as an equivalent thermal unit (or more units, one for each block of specific cost of purchase contracts) with zero start up cost, zero ramp duration, zero minimum up and down times, and appropriate input-output curves and minimum and maximum available capacity (see Fig. 6.2c and 6.2d; Fig. 6.2a and 6.2b correspond to the case of an ordinary unit).

As far as opportunity sales and secondary load are concerned, App. 2C shows how to represent them and how to consider them in conjunction with ordinary and purchase equivalent generating units.

6.2.1.2 System incremental cost grid

Fig. 6.3 represents the forecasted load curve (net from must run generation and must buy/must sell). Fig. 6.3 represents also some other curves (dashed lines) obtained by shifting the load curve up and down.

The unit commitment procedure illustrated in sec. 6.2.1.1 may be performed for each one of the curves shown in Fig. 6.3. This enables an incremental cost to be assigned for each ETI and for each load level, as identified by the various load curves. Incremental costs are calculated by the unit commitment procedure for each load level (including opportunity exchanges and secondary load) (Fig. 6.4).

If there is not enough power to cover the particular curve considered (as may happen for the curve lying over the load curve), an arbitrary, sufficiently high incremental cost will be set.

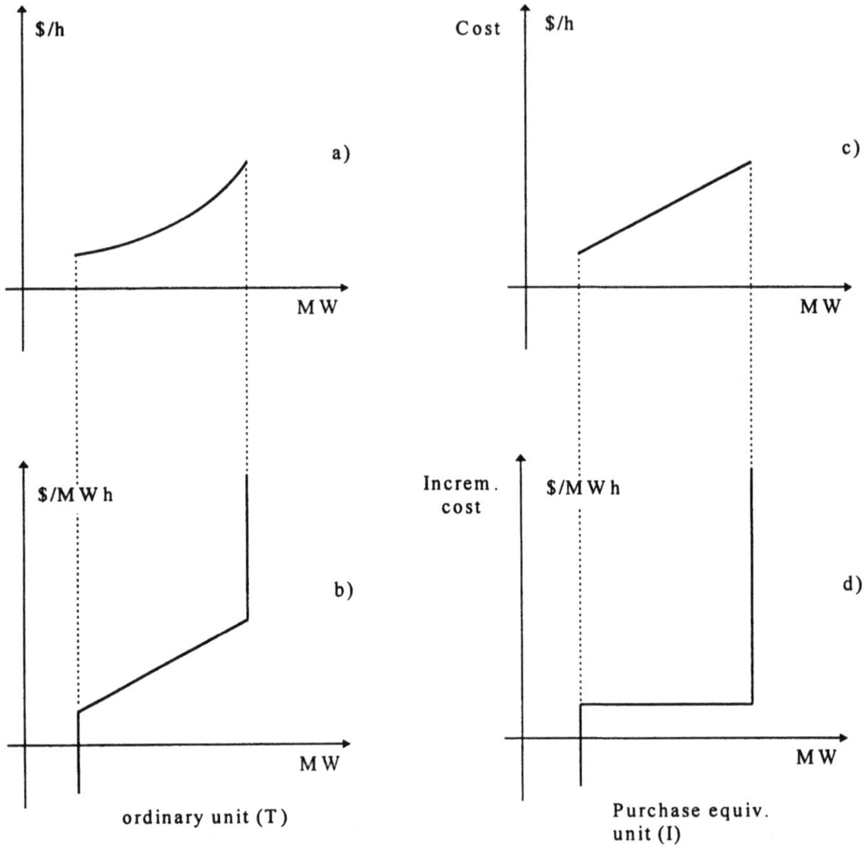

Figure 6.2: *Cost curves and incremental cost curves of a thermal generating unit (a and b) and of a particular purchase contract (c and d)*

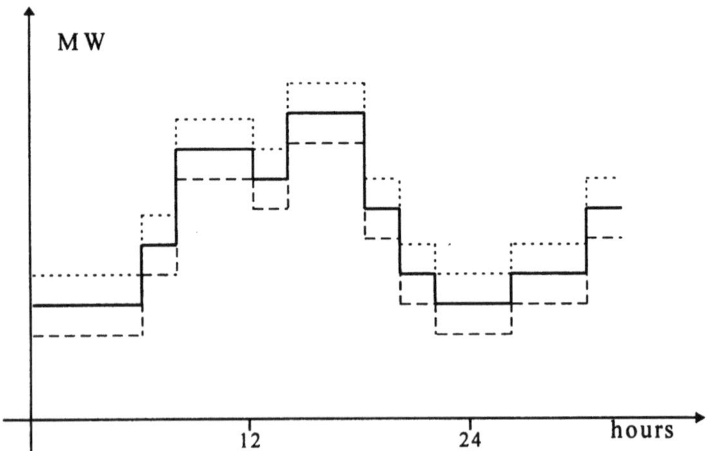

Figure 6.3: *Load curve (solid line) and shifted load curves.*

The number of curves of Fig. 6.3 should of course be limited to a minimum, due to the fact that the computation of incremental costs with the procedure illustrated in 6.2.1.1 may be time consuming; in some cases it will be possible to consider right the only one corresponding to the forecasted load curve. Nevertheless, it must be stressed that the incremental costs computed in correspondence of one given curve are valid for minor variations (up and down, around the curve itself), that is for variations not entailing an economic commitment or decommitment of thermal units.

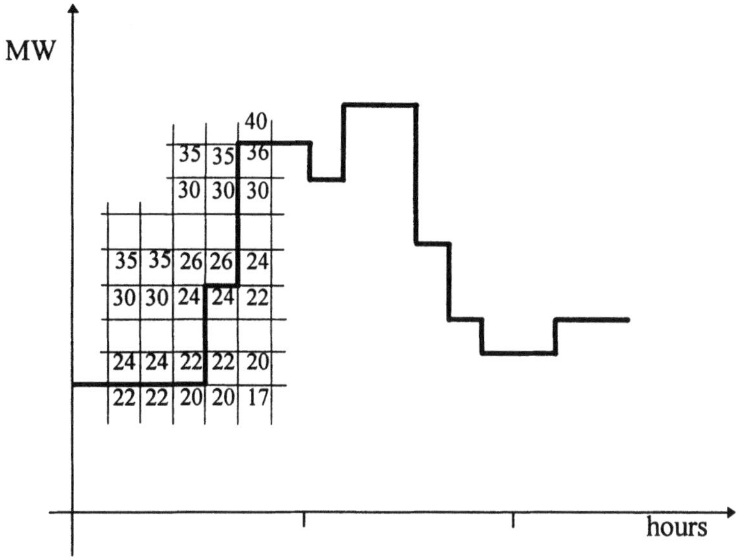

Figure 6.4: *Incremental cost grid. Solid line is the forecasted load curve. The figures shown represent the incremental costs*

6.2.1.3 Determination of hydro generation

According to definitions in App. 2A, sec 5.1, there may be some hydro generation from ponds (run of the river plants are similar to "must run" units).

Once a forecast for water inflow is available, the total output (GWh) for the period considered (one week) is known (see the remarks made about this point in App. 2A, sec 5.1).

Hydro generation will then be distributed over the week (ETI by ETI) with the criterion of covering the highest incremental cost zones of the system incremental cost grid under the load diagram.

This must be done by complying with the following constraints:

- the total (weekly GWh) output must be equal to the forecasted one, corresponding to the forecasted inflow;
- in each ETI the hydro output (MW) will be limited by lower (zero) and upper (maximum available capacity) bounds;

- the storage will be limited, at the end of each ETI, by lower (zero) and upper (maximum storage) bounds.

It may be worth noting that this methodology of allocation ("incremental costs shaving", Fig. 6.5 a), which is the optimum, normally gives results not too different from those which could be obtained with the criterion of "peak shaving" (Fig. 6.5 b), which is a near optimum solution (it can be shown that it is optimum if there are no start up or shut down of thermal units during the period considered). In both cases the efficiency of hydro power stations is considered to be independent of their MW output.

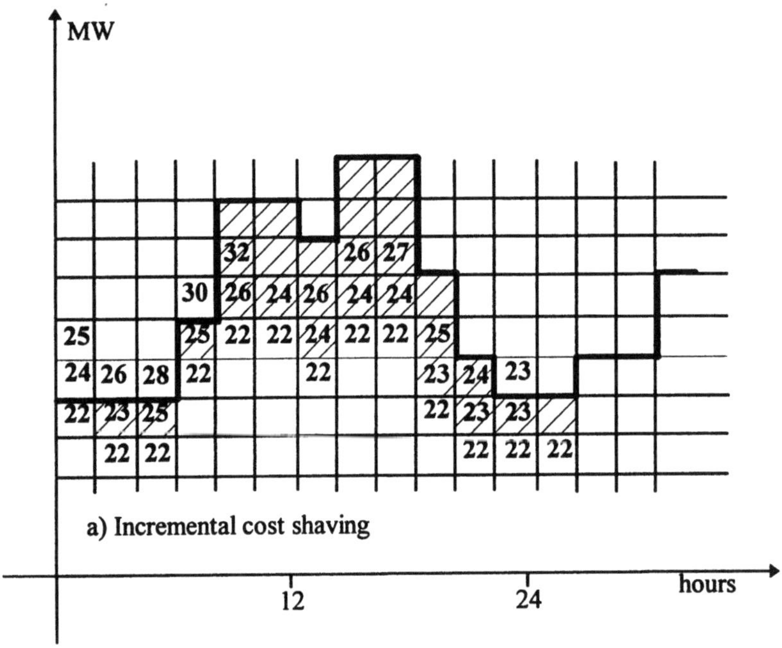

Figure 6.5a: *Allocation of hydro generation on an incremental cost shaving criterion. The two shaded areas (from 0 to 24 hours) are equal.*

6.2.1.4 Determination of generation-pumping diagrams of pumped storage power stations

The following discussion is limited to the case of pumped storage plants having no natural inflows. In the case of natural inflows a kind of superposition may be considered, splitting the plant into two fictitious power stations (with some care). A pumped storage plant may be dealt with as a normal hydro pondage plant, having a negative lower MW bound; hence the procedure illustrated in sec. 6.2.1.3 applies.

The only additional concern is that the energy balance does not check to zero, due to overall efficiency η (generation times pumping) being less than unity (normal values: 0.65 to 0.75), if water balance must check (and it must check).

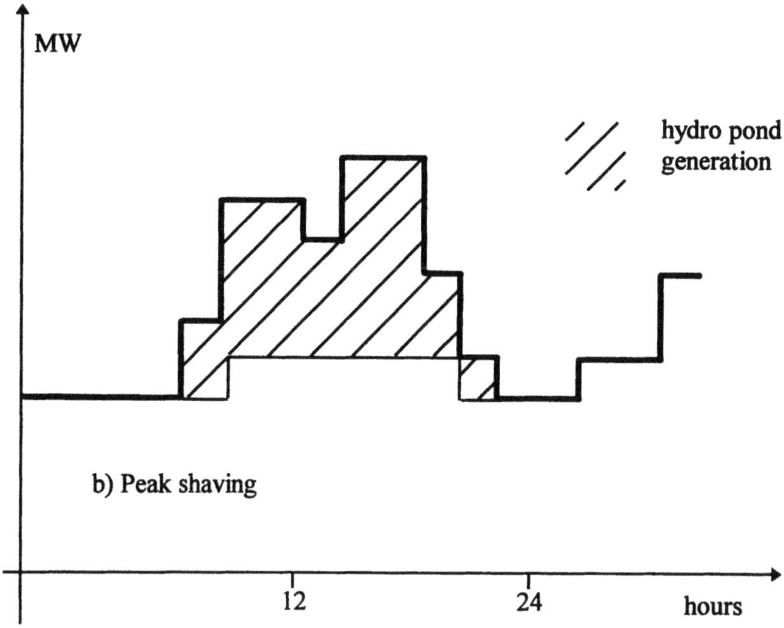

Figure 6.5b: *Allocation of hydro generation on peak shaving criterion. The two shaded areas (from 0 to 24 hours) are equal.*

The difficulty can be overcome in the following way: first determine the generation diagram of the normal hydro plants, with the procedure illustrated in sec. 6.2.1.3. Then, on the residual load curve (that is, the load curve minus the generation diagram of normal hydro plants) determine a first tentative generation pumping diagram of the pumped storage plant again with the same procedure, in which the total (weekly) GWh output is set to zero. This first tentative diagram would give rise to a negative variation in storage for pumped storage plant. As a matter of fact, p_j (MWh) being the output, positive (generation) or negative (pumping), during the jth ETI, the sum of all the P_j will be zero and the sum of the positive P_j, v, will equal the sum of the negative P_j, changed in sign; the variation in storage is then easily seen to be equal to (MWh) ηv -v, and this is a negative quantity. The first addendum corresponds to the water raised to the upper level, the second to the water turbined (which, to comply with zero total MWh output, is a quantity larger than that raised).

Then the determination of the generation pumping diagram will be repeated setting the total output at the negative value $(\eta - 1)v$.

This procedure will be repeated until zero storage variation is reached (normally it requires 2 to 3 iterations).

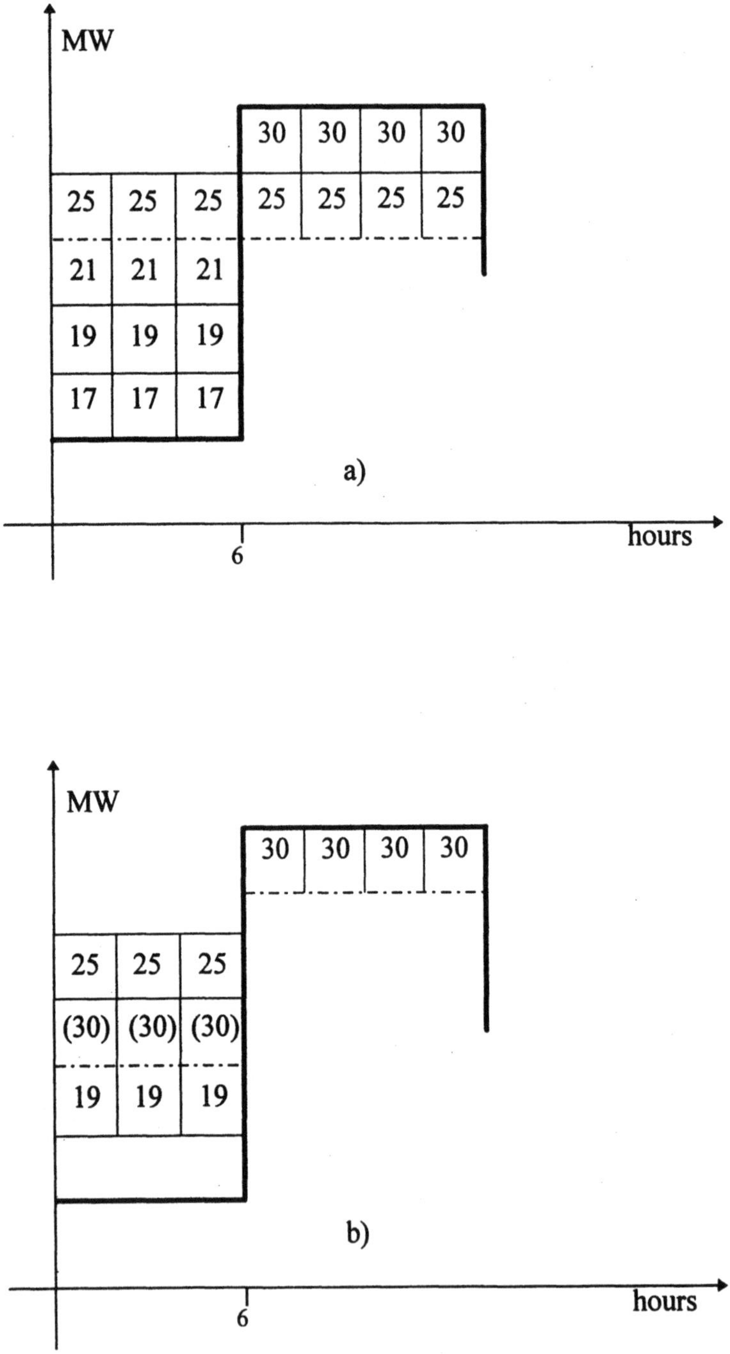

Figure 6.6: *Allocation of pumping-generation with incremental cost shaving criterion. In case (a), the pumping-generation cycle efficiency is assumed equal to 1; in case (b), equal to 0.7*

Note that, to take due account of the loss of energy $(\eta - 1)v$, when determining the generation-pumping diagram with the incremental cost shaving criterion the incremental cost read on the grid must be increased by dividing it by η, whenever p_j is negative.

Fig. 6.6 a represents the original incremental cost grid, the load curve (net from hydro pond generation) and a possible (dashed line) generation-pumping diagram; Fig. 6.6 b shows the modified incremental costs (assuming $\eta = 0.7$), and the resulting generation-pumping diagram.

Note that the procedure described here considers only the "short term energy transfer" duty of pumped storage plants.

6.2.1.5 Determination of thermal generation diagrams

If medium-long term operational planning exists, the resulting values of T, I, E and L must be complied with (these are respectively the free thermal generation, the secondary import and export and the secondary load, see App. 2A).

The full procedure to follow in the more general case is shown in Fig. 6.7.

Note that the determination of hydro pond generation schedule and generation-pumping schedule is iterative, as well as the determination of system incremental costs grid, because they are interdependent. The cost grid depends on the diagram to be covered by generation of thermal units (plus the other thermal-equivalent items), and this diagram is the difference between the load (network demand) diagram, which is known, and the hydro pond generation plus generation-pumping diagram of pumped storage plants.

So, at least two iterations are necessary. The "first tentative" incremental cost grid (which is produced just at the beginning of the scheme of Fig. 6.7) is determined with reference to the load curve, as shown in Fig. 6.3 and Fig 6.4, since the hydro generation diagram is not yet known at this point ; the "second tentative", on the contrary, is determined, as it must be, with reference to the load curve minus hydro pond and generation-pumping diagrams, that is, the "residual curve no. 2" of Fig. 6.7, which is the right diagram to be covered by thermal units.

As an interesting point it may be noted that in the second iteration the pumped storage plants could displace a previously selected (committed) thermal unit.

Often the procedure will not be so complex as it appears in Fig. 6.7, due to the absence of some items such as pond hydro, or secondary load, etc.

For example, the scheme of Fig. 6.9 corresponds to the case with no constraints from medium term scheduling and no hydro pond generation nor pumped storage: Fig.6.10 to no hydro pond nor pumped storage.

6.2.1.6 Split saving

It often happens especially with thermal utilities of small to medium size that they cluster in pools in order to take advantage of co-ordinated operation.

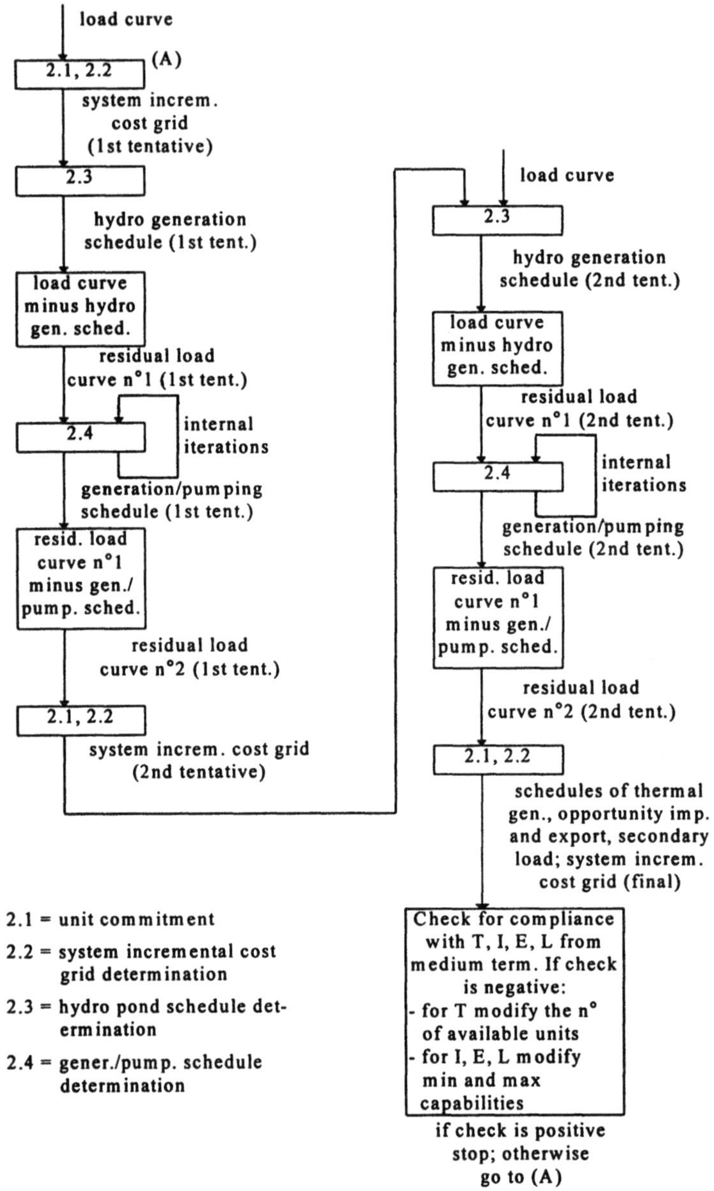

Figure 6.7: *Overall short-term scheduling procedure*

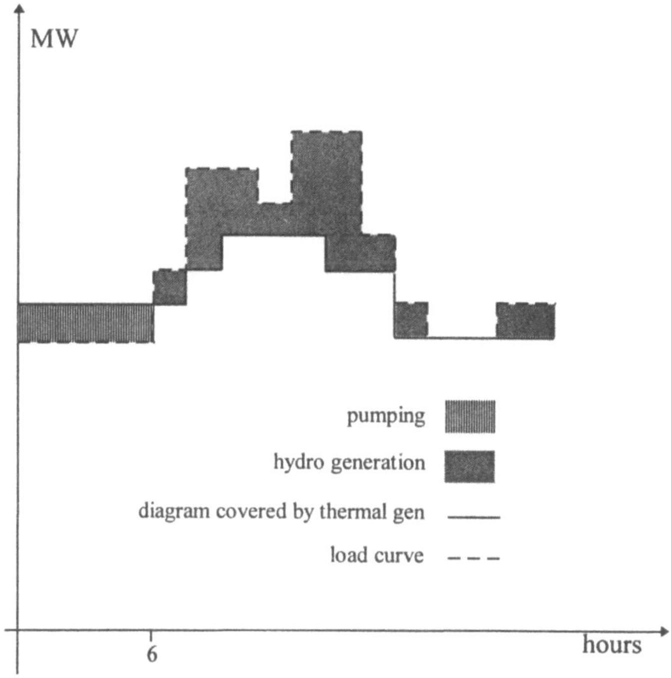

Figure 6.8: *Coverage of a load curve by hydro plants*

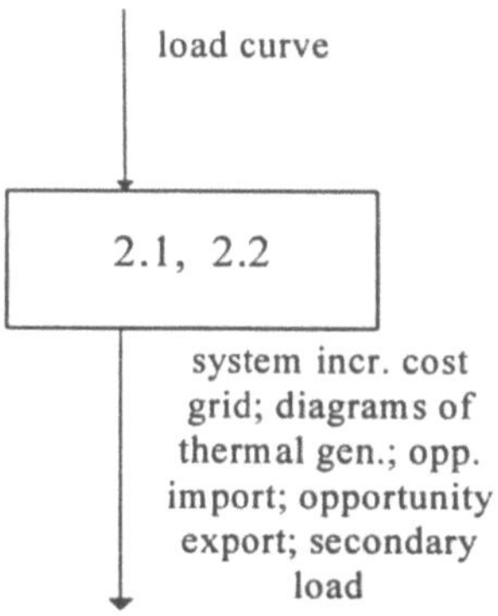

Figure 6.9: *Short-term scheduling procedure in case of neither medium-term scheduling constraints, nor hydro generation (compare with figure 6.7)*

Suppose that a co-ordinating body performs the procedures described in sec. 6.2.1.1 to 6.2.1.5, as appropriate, applying them to the whole of the k interconnected systems. The optimum schedule will entail operation (weekly) costs C_1, C_2, ..., C_k which must be carried by the participating utilities, so that

$$C = C_1 + C_2 + \ldots + C_k = \min.$$

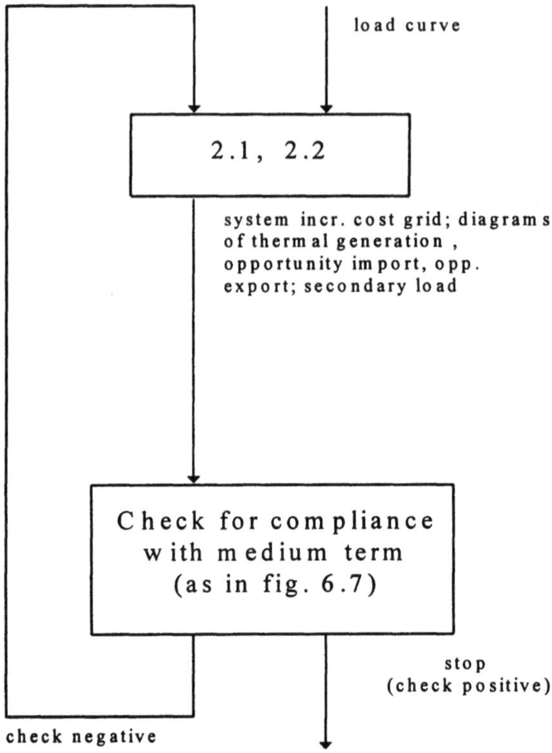

Figure 6.10: *Short-term scheduling procedure in case of no hydro generation (Compare with figure 6.7)*

Of course, there will be power exchanges between the various utilities, and let P_i be the export (GWh) of utility i towards the rest of the k-1 utilities (P_i may be negative, and $P_1 + P_2 + ... + P_k = 0$).

Suppose now that the same procedures 6.2.1.1 to 6.2.1.5 are applied separately to each one of the k utilities, these supposedly not being interconnected (or perhaps having separate, firm trading agreements, which should be known).

The optimum costs will now be C_1', C_2',.......,C_k' and $C' = C'_1+C_2'+......+C_k'$ will be greater than C.

The saving implemented with interconnected and co-ordinated operation is then

$$C' - C.$$

The split of this saving between the utilities may follow a variety of criteria. In one practical case the split is as follows:

- the co-ordinating body receives 0.1 (C' -C)
- each utility receives $0.45(C'-C)L_i / (L_1+L_2+..........+L_k)$, L_i being the firm load (GWh) of utility i
- each utility receives furthermore

$$1.1 \cdot 0.45 \cdot \frac{P_i}{p_1 + p_2 + \ldots + p_k}$$

if P is positive (exporting utility; p is the absolute value of P), or

$$0.9 \cdot 0.45 \cdot (C'-C) \cdot \frac{-P_i}{p_1 + p_2 + \ldots + p_k}$$

if P_i is negative (importing utility).

All the expressions are positive, and it can easily be seen that their sum equals C'-C.

In this case a penalty is imposed on the importing utilities (to encourage the building up of new plants).

Further penalties, not illustrated here, are in some cases applied to those utilities which are not able to meet their own peak load with their own plants.

Of course, in addition to the split saving illustrated above, the power (energy) exchanges will be settled on the basis of bilateral contracts.

6.2.2 Hydro power systems

Hydro power systems are those with medium-long term storages.

The existing medium-term scheduling procedures may be grouped in the following categories, as far as short-term hydro scheduling is concerned:

1) - of the type of Sec. 2A.5.2, with aggregated reservoir ; data transmitted to the short-term are H, I, L, E (GWh) for the coming week, that is, hydro generation from reservoirs, secondary imports, load and secondary exports;
2) - of the type of Sec. 2A.6.2, with aggregated reservoir; data transmitted to the short-term are the water value w of the coming week, this being considered a more stable value than H, I, L, and E;
3) - as in the case of 1), but with disaggregated reservoirs; the data transmitted to the short-term are $H_1, H_2, \ldots, H_N$, I, L, E (GWh), N being the number of reservoirs;
4) - as in the case 2), but with disaggregated reservoirs: $w_1, w_2, \ldots, w_N$.

In cases 1) and 2) a disaggregation is anyway necessary. If the hydrologies of the various watersheds are similar, a simple proportionality of the per unit storage may apply; if this is not the case, the disaggregation should anyway have taken place in a previous phase.

Hence it may be supposed that the input to short-term procedure is in any case given in the form 3) or 4) above.

It should be noted also that, in addition to hydro reservoir generation, at least one of the following items is present: opportunity import I, opportunity export E, secondary load L.

At the end of sec. 6.2.1.1 it was shown that import, export and secondary load might not be considered as fictitious or equivalent thermal units (Fig. 6.2 and Appendix 2C). This allows to build up, also in the case of hydro power systems, an incremental cost grid like the one discussed in Sec.6.2.1.2 and shown in Fig. 6.4.

Nevertheless, in this case it is likely that, unlike Fig.6.4, the zone covered by real incremental costs would be at the bottom, due to the reduced value of opportunity items (I, E, L) with respect to hydro reservoir. The upper portion of the load curve, will be filled by arbitrary, sufficiently high incremental cost values (see Sec.6.2.1.2) (Fig.6.11).

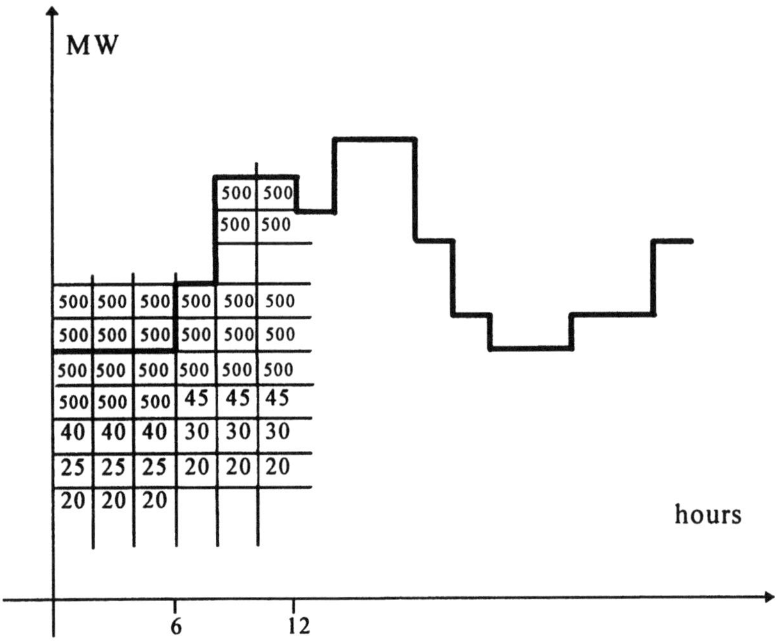

Figure 6.11: *Incremental cost grid in case of hydro dominant system. The upper part is filled with fictitious arbitrarily high values*

6.2.2.1 Assigned reservoir generations

This corresponds to case 3) above. The scheduling of course interests the ponds also.

The allocation of reservoir and pond generation over the week (ETI by ETI) may be done with the criterion of incremental costs shaving (as in the case of Fig. 6.5 a) or, with some approximation, with the criterion of peak shaving (Fig.6.5b).

In both cases various constraints must be taken into account; some of them are like those considered in Sec. 2A.4 (max. power of each power station, upper limit of storage of each pond and reservoir); others are more specific to the short-term problem, due to short duration of the ETI and to the detail (plant by plant); with which the scheduling must be set up, namely, influencing and water delays between the plants.

All these constraints may be dealt with by means of particular algorithms, like network flow techniques, which are not proposed to be discussed here because of the mathematical nature of the problem.

Some account of the best use of water may be introduced in a simple manner, by inhibiting running at output less than predetermined values (e.g., not less than 10% for Pelton turbines, or 30% for Francis), that is, at too low values of efficiency.

6.2.2.2 Assigned medium-term water values

This corresponds to the case 4) above.

The water values w_1, w_2, ..., w_N will be considered constant all through the week.

Note that the water value is defined in medium-term operation planning only for reservoirs (not for ponds).

While in the case of Assigned reservoir generations 6.2.2.1 the total GWh to be generated from reservoirs and ponds are both assigned, in this case only the GWh from ponds are assigned (on the basis of water inflow forecast).

The short term schedules may be defined in this case by considering the hydro reservoir power stations as equivalent thermal units (having zero start-up cost etc.), with an incremental cost equal to w_i (or in some way weighted with efficiency, as in (18) of App.2A); hence the two procedures 6.2.1.1 (unit commitment) and 6.2.1.2 (incremental cost grid determination) may be performed. Hydro pond generation schedule could then be determined with the incremental cost shaving criterion as illustrated above (sections 6.2.1.3 and 6.2.1.4).

6.2.3 Hydro-Thermal Power Systems.

In practice the algorithms illustrated in sec.6.2.1 and 6.2.2 apply to this case as well.

Referring for example to Fig. 6.7, the only modifications required are:

I. if the input to short term from medium term, as far as hydro reservoirs are concerned, is constituted by the generations H_1, H_2, .., H_N (see sec.6.2.2, iii), then the reservoir power stations may be regarded as ponds, and Fig. 6.7 fully applies;

II. if the input to short term from medium term, as far as hydro reservoirs are concerned, is constituted by the water values w_1, w_2,.., w_N (see sec.6.2.2, iv), then the reservoir power stations may be regarded as equivalent thermal units, and Fig. 6.7 again fully applies.

6.2.4 A new framework of scheduling procedures

Fig. 6.7 shows an iterative procedure in which unit commitment, hydro generation, thermal generation and opportunity trading determination, etc., are interleaved.

Some utilities are beginning to test (in some cases to utilise) a different frame, in which there are two hierarchical levels; the upper level is a "co-ordinator", the lower level is constituted by some "optimisers", one for each source or item of MW balance.

Fig. 6.12 gives an idea of this new structure. In the initial phase the co-ordinator gives a series (one value for each ETI) of system incremental values or costs to the optimisers.

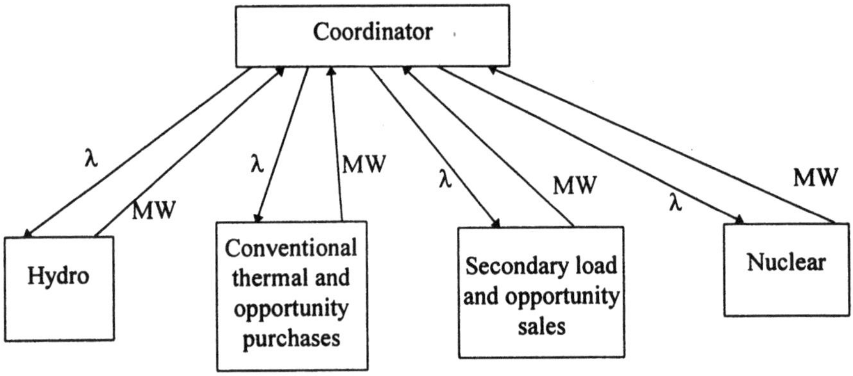

Figure 6.12: *A new framework of scheduling procedure*

At the first iteration each optimiser, independently of the other, determines an optimal schedule of its own item, taking into account the incremental values received and the local constraints (these are, for example, for the hydro sub-system, the amounts of water to turbine, the links between the plants, etc.).

The co-ordinator compares, ETI by ETI, the MW totals, received by the optimisers, with the load; if the balance does not check, it modifies in appropriate manner the incremental values (raise if there is a surplus, diminish if there is a deficit) and transmit them back to the optimisers.

Then the second iteration takes place, and so forth.

The results do not differ from those obtainable with the more conventional procedure of Fig. 6.7. Which of the two methods is better suited, especially in the case of big power systems, is a matter of available mathematical algorithms.

For example, the way the incremental values are modified by the co-ordinator is important with respect to convergence.

6.2.5 Conclusions.

Each utility develops or adopts a short-term scheduling procedure to meet its own requirements. These procedures generally aim at selecting the generating plants and operation to meet expected demands and trading requirements as economically as possible according to an acceptable standard of security. Procedures range from using simple merit order listings to sophisticated computer programs. Some simplifications in the representation of the power system (especially hydro subsystem) seem to be anyway necessary. They must be carefully chosen when building up the algorithm.

The conflicting requirements to be met are mainly:

- sufficient accuracy in the determination of optimal and secure schedules (in some cases even feasibility may be difficult to achieve, for some constraints are difficult to represent);
- robustness of the procedure: a long testing time may be required to make sure that all particular cases are taken care of (the simplifications adopted are in general the main cause of failure of convergence to optimal solution, or of yielding unfeasible solutions);
- acceptable execution times and computer memory requirements.

The procedure should also have good man-machine interfaces, to make possible easy modification of input data (e.g. changes in load, water inflows, spot market opportunity), control of intermediate results and saving, etc.; this is of particular importance in the cases in which the scheduling procedures have to be managed also by the control room staff.

Improvements are needed, in particular in the field of load forecast and plant availability forecast.

Interfaces with longer term scheduling and on-line operation should be provided.

Skill of scheduler staff is still of great importance, even when sound and established procedures are available.

Especially in the case of complex power systems, an easy update and amendment of databases is mandatory; this requires a careful structuring of the various software blocks which constitute the whole procedure.

Some allowance should also be made for substitution of individual software blocks, when more efficient mathematical methods of solution become available.

6.3 Medium-long term operation planning

To facilitate a description of the methodologies adopted to produce medium-long term generation schedules or strategies, it is useful to clarify concepts such as, for example, water value, cost curves, firm- non-firm energy, cost or value of unsupplied energy, etc.; section 6.3.1 is devoted to such a description.

Section 6.3.2 deals with a specific aspect, characteristic of medium-long term operations planning, that is, on how one can, in principle, account for randomness affecting the following items: load, energy exchanges with other utilities, availability of generating units, energy availability (e.g., water inflows).

Section 6.3.3 illustrates the methodologies adopted by some utilities; the utilities have been chosen in such a way as to cover, as far as possible, a variety of sizes and generation mix; a limit to the choice of examples was nevertheless put by the fact that not all utilities have a medium-long term operation planning procedure in routine operation.

Section 6.3.4 gives some conclusions.

It should be stressed that in the field of medium-long term operation planning the research work is still in progress, and that some problems have not yet been satisfactorily solved by all the concerned utilities: for example, the co-ordination of two or more interconnected areas with different hydrologies, or a more appropriate modelling of the reliability of thermal units. References will be made to such problems during the illustration of the particular procedures.

It should be mentioned that medium-long term operation planning has mainly a task of decision support; that is, it is seldom possible to use automatically its results as input to shorter term scheduling. This is due not only to the presence of various uncertainties and of random values, but also to the many simplifications adopted.

6.3.1 Definition of some key concepts

6.3.1.1 Value of reliability

It is well known that in most electricity supply acts of the State-owned electric utilities (and perhaps also in some private-owned ones) it is stated that the electric service should be *economic* and *reliable* (and, more recently, have concern for the environment); this two-fold goal may be stated explicitly in two equivalent (but formally different) ways (2):

1) minimise the operation costs, subject to some *constraints* which guarantee a predetermined level of reliability (e.g. spinning reserve MW higher than or equal to some preassigned value; expected unserved energy MWh - EUSE- lower than some per cent of energy demand; loss of load probability -LOLP- less than a predetermined figure; etc.);
2) assign a cost to unserved energy (\$/ kWh), then *minimise* the operation costs, including the cost associated with expected unserved energy (the last being considered a generation-like resource, characterised by the above-mentioned cost).

The first method is the one which is more widely followed, even if it seems, from a theoretical point of view, less sound than the second one; as a matter of fact, the limits of spinning reserve or of EUSE are determined on a heuristic basis, following past experience and perhaps more subjective assumption of the management.

The second method appears more appropriate: should the cost of USE (\$/kWh) be known, the limits of spinning reserve (MW) or of EUSE (MWh) would be a result of the optimisation (cost minimisation) procedure.

Unfortunately, it is difficult to give a definition of the cost of unserved energy; in fact, it is easily understood that it depends on various factors which are well beyond the reach of experience and direct responsibility of an electric utility ("external, non-monetary costs"); moreover, it is likely to change in time according to unforecasted events, modifications in social structure, etc.

Nevertheless, some attempts have been made in the past to quantify the cost of unserved energy (only direct, "internal or monetary costs"); one approach that has been employed is the *production factor analysis* [3].

It assigns to the unsupplied kWh the value of the goods produced through it, including damage and clean-up expenses of the production facility. In this approach it is necessary, besides other factors, to account for the duration of the interruption (which can have influence on damages and also on the amount of lost production); the values so determined for the various consumers shall then be averaged; as a final result a cost of unserved kWh is found, which shall be in general a function of time of the year (of course, the estimate is done with reference to the daily peak; the variation in time is likely to depend on the season of the year); it will be also a function of the duration (e.g., a few minutes to hours) of the interruption.

Customer surveys are used to help the evaluation of the above-defined costs, including perhaps some other components referring to external or non-monetary costs.

The definition of the cost of unsupplied energy is subject to some uncertainty and variability, mainly due to different mix of consumers served. The value most frequently recognised is around 1$/kWh, but the range, from utility to utility, is as wide as 0.5 - 5 $/kWh or more.

As far as the first method [1] above is concerned, a spinning reserve equal to the largest generating unit synchronised to the system is often assumed as the operating rule [1].

One utility has assumed an upper limit of EUSE equal to 10 system-minutes per year (firm load; that is, this value does not include interruptible consumers).The unit "system-minute" is defined as the amount of energy obtained by multiplying the peak load of the system, in the year considered, by one minute; e.g., for a system having a peak load of 30,000 MW, 1 system-minute corresponds to 500 MWh, and 10 system-minutes to 5,000 MWh.

With a load factor of 0.6 (that is, yearly load demand equal to 0.6 times the peak load times 8760 hrs), 10 system-minutes correspond to about 0.00003, or 0.003%, of yearly load demand.

Interest in considering EUSE as reference parameter for security is shown mainly by hydro-dominant utilities; other utilities often prefer to use LOLP - loss of load probability.

Lastly, it may be interesting to note that method [2] above could be adopted, at least from a methodological point of view, assigning to EUSE a conventional cost, determined on the basis of the experience and depending on higher or lower availability of reserves (that is depending also on the criteria of method [1]). In fact, computation algorithms are in general more straightforward, from a conceptual point of view, in case [2] (the curtailment of demand may be simply represented by an equivalent generating unit of conventionally high cost) [2], [6], [7]. In the planning phase sometimes the value of USE (average in the year) is assumed equal to the ratio between GNP and total electric energy demand.

6.3.1.2 Firm and non-firm energy

This concept refers to system load, energy exchanges with interconnected utilities, and generation. Substantially, a block or parcel is said to be "firm" if it is contracted or committed without possibilities of modifications. A more detailed definition is given in Appendix 2A, sec 3.

6.3.1.3 Cost and revenue functions or curves

Cost functions refer to thermal generation and secondary energy imports; revenue functions, to secondary export and secondary load.

A cost or revenue function refers to a given time interval; having known the various amounts of thermal generation, secondary load and secondary imports-exports (GWh) available, an optimisation within the considered time interval is performed, so that for each amount of generated or purchased energy the corresponding expenditure is minimum, and for each amount of sold energy the income is maximum. Appendices 2A, sec. 7, and 2B illustrate in some detail the definition and possible use of these functions.

6.3.1.4 Elementary time interval (or time stage) in medium-long term operation planning.

The period under study is divided into a number of "sub-periods" or "elementary time intervals" (for example, a 3-year period may be divided into 36 sub-periods, each of one month duration). In each elementary time interval all the variables of the problem (e.g. thermal generation, load, hydro generation, etc.) are considered constant with respect to time; in other words, an elementary time interval is a discrete point along the time axis. Appendix 2A.1 gives a description of possible time partitioning in operation planning problems, and of the assumptions underlying the choice.

6.3.1.5 Value of water

This concept is especially important for hydro dominant and hydro-thermal utilities.

A "medium term water value" w is defined, as a function of time, in Appendix 2A, sec 6; this is limited to the case of deterministic approach of medium-long term scheduling.

In section 6.3.3.1 and Appendix 2D some considerations are developed for the stochastic case.

6.3.2 Operations planning in randomness

Three main factors are typically subject to random variations, namely, the load, availability of the generating units, and availability of the primary energy (water inflows to hydro plants).

The "random" character of the above-mentioned variables unlike the "uncertainty" of others such as the future prices of fuel or the currency exchange rate with fuel-exporting countries. Uncertainty - as opposed to randomness - is often dealt with by performing sensitivity analysis.

When the period of interest is of short duration and concerns the near future (up to 1 to 10 days ahead), the forecast of those random factors is in general aleatory within a limited degree, so that it is possible, or acceptable, to consider them as deterministic (through the aid of well-known procedures, such as load forecast and water inflow forecast techniques); the most probable values may be assumed, and the allocation of appropriate (and easily computed) reserves is sufficient to face expected deviations.

In this context, deterministic optimisation procedures may be used (see sec. 6.2 and ref. [1]).

As a result, generation schedules are drawn up representing a more or less strict guidance to shift operators (a schedule is a sequence of values, e.g. of thermal generation, one for each elementary time interval of the period under study).

On the contrary, when the period of interest is of longer duration, it is not possible to assume the deterministic forecast; the randomness of the main factors mentioned above should be accounted for in a more appropriate fashion.

In order to fix the ideas, let us suppose that the period under consideration covers the next 36 months (starting from the beginning of the next month of the calendar), and that the elementary time interval is one month.

It is supposed that the dates on which new plants are commissioned are known; that a procedure to produce sequences of future, monthly water inflows in a probabilistic manner is available (accounting, if it is the case, for co-relation between two adjacent elementary time intervals); that reliability parameters of existing and future generating units are known (as well as their maintenance schedules); that a load (monthly energy) forecasting procedure is available, giving the most probable values and expected deviations, or similar parameters; and that the forecast of market, or secondary load, availabilities are sufficiently accurate.

6.3.2.1 Operation strategy or closed loop.

One methodology, often referred to as operation strategy or "if-then" or closed-loop methodology, consists in producing, normally through the use of algorithms based on stochastic dynamic programming (SDP), the so called decision tables.

A decision table is a matrix that, for each month, gives the value of the control variable (for example, hydro generation, GWh) which minimises the expected (future) cost; the optimal value of the control variable is given as a function of the present state of the system, that is, typically, of the amount of storage in the aggregated hydro reservoir, and of the amount of water inflow which presented itself in the month just elapsed (this is a state variable if the assumption of correlation between adjacent time intervals was assumed; otherwise, the state is represented by storage only) (Fig. 6.13).

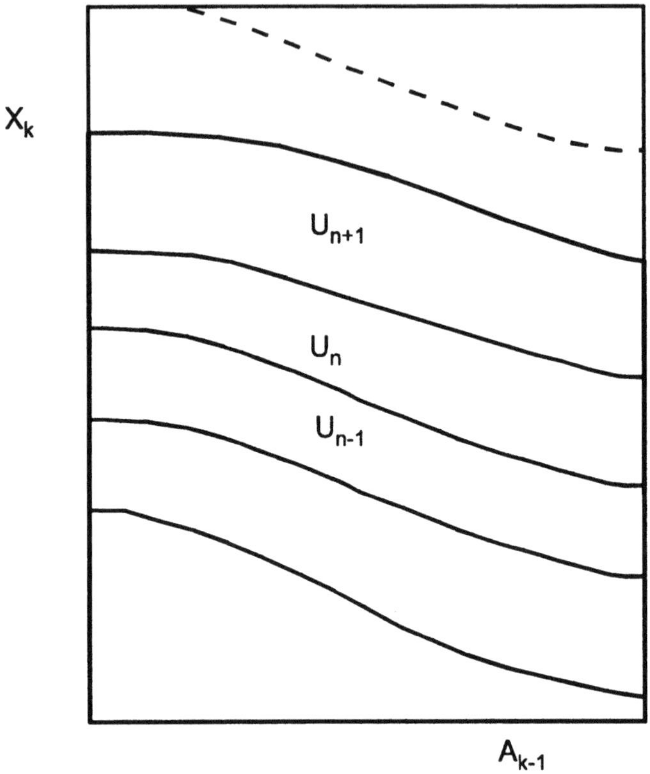

Figure 6.13: *Decision table for k-th time interval. Ordinate: Storage X_K at the beginning of k-th time interval. Abscissae: water inflow A(k-1) during (k-1)th time interval. The lines shown are equi-U curves (U being the control variable, e.g. the water turbined in the kth time interval). U(n+1)>U(n) (Ref. [5])*

This methodology, hence, does not produce schedules, because the optimal value of control variable for month k+1 becomes known only when the two state variables (storage and inflow) are known, that is only when month k has elapsed; in other words, it gives operating results only for the next elementary time interval.

On the other hand, it is obvious that when considering the random future, schedules may not be produced; the decision of generation can refer only to the next time interval, provided that all the hazards and decisions of the past have led to a given state, assumed to be known.

The tables are produced, when the computation is performed, for each one of the 36 elementary time intervals or months; but each table becomes usable only when "activated" by the calendar.

The need to consider all the future period (36 months, in the example considered) should then be a matter of concern, if only one month (the next) is the object of a result or decision, as stated above.

Yet, that need is an obvious requirement for the optimisation; theoretically, the period to be considered should be long enough to make it possible to assign a negligible value to the final storage (taking into account an appropriate discount factor, which transforms future values into present ones).

The computation should be repeated, covering again the next 36 months, whenever a major variation occurs in the parameters which are assumed as known (commissioning of new plants, maintenance schedules, reliability parameters of generating units, etc.); on the contrary, it needs no re-execution depending upon random variations (within the limits of the historical series) of load, availability of generators, and water inflows.

This kind of algorithm is quite demanding in terms of computing power and time; hence the hydro system is normally represented by a unique, aggregated or equivalent reservoir; to have n (equivalent) reservoirs represented separately would increase the number of state variables from 2 to 2n. The limitation of a unique reservoir is considered acceptable if the hydrologies of the various watersheds are similar; nevertheless, in this case also, the control of the risk of spillage or of depletion of some particular reservoir can get lost or unsatisfactory.

Some utilities are developing algorithms which could allow the representation of 2 or 3 equivalent reservoirs (allowing also a control of energy flows between the corresponding electrical subnetworks) without a prohibitive increase in computing time and power requirements; in some cases (through the use of "aggregation-disaggregation" methods) the number of equivalent reservoirs could reach 10 to 15 (the computing time, in principle, increases only linearly with the number of reservoirs, when using these techniques); these methodologies are illustrated in [2], [4], [5], [6]. Even worst is the case in which electrical subsystems are also hydraulically coupled.

6.3.2.2 Operation policy or open loop

Another methodology, referred to as operation policy or open loop, is more or less similar to that used in short term, that is deterministic, but devotes more attention to the definition of the constraints; the aim of this more appropriate definition of constraints is to avoid, within reasonable probability limits, curtailments of load or water spillages.

The constraints may be, typically, minimum and maximum storages in hydro reservoirs, minimum amount of pumped energy (which may be required to comply with minimum storage requirement), etc.

Those constraints are then used in a deterministic optimisation procedure, in which the random parameters (load, water inflows, availability of generating units) are set at their most probable, forecasted values; as a result schedules of hydro and thermal generations are produced, in some cases also with confidence limits. This allows to make an estimation, for example, of future needs of fuel and hence to provide purchase schedules (contracts); this is not the case, of course, with the methodology illustrated in section 6.3.2.1

This is perhaps the reason for the preference given to the methodology 6.3.2.1 by hydro dominant utilities, and to methodology 6.3.2.2 by thermal dominant utilities. A purely thermal utility will not, of course, make use of methodology

6.3.2.1. For a purely hydro utility the methodology 6.3.2.1 appears to be preferable (the optimisation concerns in this case the secondary markets).

The operation policy procedure is in general repeated each month, covering always the next 36 months; this allows to better account for random events, as they become nearer in time and hence easier to forecast (this is also called an open-closed loop procedure).

6.3.2.3 Deterministic computations

The presence of random values may be taken into account in a rigorous way through methodologies 6.3.2.1 or, to some extent, 6.3.2.2; "rigorous" has here reference only to the aspect of randomness, and does not imply the absence of simplifications in the representation of the power system (these simplifications are on the contrary present in most cases: e.g. aggregated instead of detailed hydro system). The presence of random values in a "rigorous" way strongly affects the operation planning procedures, making them quite complex. This is the reason why many utilities (especially thermal dominant) prefer to work with average forecasted values (deterministic computations), and simply consider, to account in some way for randomness, an additional (but limited) number of alternatives for the most important random parameters (global values), such as water inflows and load (sensitivity studies). This choice is supported also by the fact that the results of medium-long term operation planning procedures have in most cases, even when methodologies 6.3.2.1 or 6.3.2.2 are adopted, only the task of decision support, and not that of giving precise or mandatory input to shorter term scheduling.

6.3.2.4 Comparison of closed loop and open loop methodologies

A comparison of the two approaches 6.3.2.1 and 6.3.2.2 has been tried by one hydro dominant utility, in order to determine the expected magnitude of saving obtainable with the more appropriate approach 6.3.2.1.

The comparison was performed as follows.

1) Compute, through methodology 6.3.2.1, the 36 decision tables;
2) sort out one sequence of inflows (36 values); with the aid of decision tables determine the schedule corresponding to the sorted inflow sequence (that is, make a "simulation" of the operation), and compute the corresponding total operation cost;
3) repeat the step 2) an appropriate number of times (e.g., 50), of course with different inflow sequences;
4) compute the mean value c_r of the 50 computed values of operation costs, and its standard deviation σ_r;
5) with each one of the 50 inflow sequences above sorted out, make a deterministic optimisation through methodology 6.3.2.2; the total operation cost is then known for each one of the 50 cases;
6) compute the mean value c_d of the 50 computed values of cost of step 5), and its standard deviation σ_d.

In the example considered, assuming as reference value $c_r = 1$ and $\sigma_r = 0.02$ (this being the result of the methodology 6.3.2.1 plus simulations), the result of steps 5 and 6 was $c_d = 1.03$ and $\sigma_d = 0.07$. Hence the expected saving is 3%, in the example (the amount of energy or of money to which the 3% applies is in this case that corresponding to the secondary market, which in turn represents a relatively small amount of the total energy balance).

6.3.2.5 Conclusion

Whatever be the methodology used, hydro subsystem needs a global (only one equivalent reservoir) or partial aggregation. The aggregation, depending especially on mutual influencing between plants, if any, may not be simple to perform. That is, the equivalent reservoir power station in general may not be simply represented by the sum of installed powers, of maximum and minimum storages and of water inflows of the actual reservoir power stations. A method of getting the parameters of the equivalent reservoir power stations is illustrated in ref. [10]; the method is based on simulations of daily operation of the detailed hydro system, under an appropriate number of power availability and water inflow hypotheses.

6.3.3 Methodologies in use in some utilities

The major purposes of setting up a medium-long term operation planning procedure are with a view to producing generation schedules (also when "strategy" methodologies are used as a first step); checking generation maintenance schedules; determining fuel requirements for setting up purchase contracts to produce input for shorter term generation scheduling to check the need for new generating plants (aid to expansion planning).

The main improvement sought over the existing procedures is a better load forecast.

The generation plans or schedules are updated periodically, or upon major outages, or upon important variations in power exchange opportunities with other utilities,

Key for identification of utilities:

A = Oslo Electricity Board

B = Hydro Quebec

C = Centrais Electricas Brasileiras

D = ENEL (Italy)

E = Ontario Hydro

F = Swedish State Power Board

G = Electricte de France

H = Iberduero S.A.

L = CPTE (Belgium)

M = CEGB (Great Britain).

6.3.3.1 Purely hydro (thermal energy generation less than 15% of the load)

Three utilities (A, B, C) have been considered. Utility C has some thermal generation, and the other two have secondary market opportunities; nevertheless, thermal generation and market opportunities represent a minor component of the energy balance.

The methodology used is that outlined in section 6.3.2.1, with some simplifications consisting mainly in considering deterministic load (monthly energies) forecast and thermal (or opportunity markets) availabilities; water inflows are supposed to be random, with specified stochastic law (derived from historical sequences).

Let X_k be the storage of the equivalent reservoir (GWh) at the beginning of time stage (or month) k; U_k the hydro generation from reservoir during month k; $F_k(U_k)$ the revenue, or monthly profit, function (Fig. 6.14), obtained as shown in sec. A.7. of Appendix 2. This function includes, due to the preoptimisation method used within the elementary time interval k considered, the effects of all the items which are significant in respect of costs and revenues, namely, thermal generation, purchase-sale opportunities and, if necessary, load disconnection; in other words, for each value of U_k, corresponding (optimal) values of thermal generation, opportunity purchase and opportunity sale are predefined (see A.7 and B).

The object function is then

$$\max_{U_k, A_k} E\left(\sum_{k=1}^{N} \gamma^k F_k(U_k) + \gamma^{N+1} I\right)$$

subject to well-known constraints (upper and lower limits for U_k and X_k, etc.). In the preceding expression E is a mathematical operator which represents the expected value (that is the average) of the expression in the brackets, taken with respect to the variables A_k (water inflow during month k), for all values of k from 1 to N (this being the total number of months constituting the period considered, that is 36 in the example of section 6.3.2.1); γ is the monthly discount factor; I is the terminal value function, which is a function of final storage $X_{(N+1)}$ and may be to some extent of an arbitrary shape, but monotonically increasing with $X_{(N+1)}$ and with decreasing derivative.

Another position of the problem, with recursive formulation, is

$$I_k = \max_{U_k, A_k} E\left(F_k(U_k) + \gamma I_{k+1}\right)$$

with k varying from N to 1 (backwards).

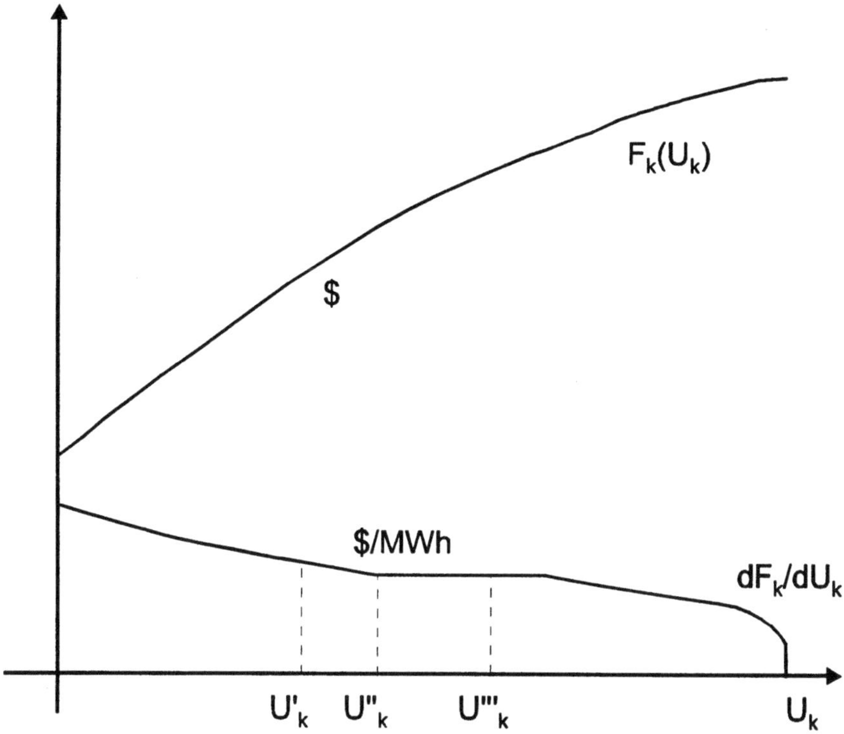

Figure 6.14: *$F_K(U_K)$ curve and its derivative. U_K', U_K'', U_K''' correspond respectively to A_{k-1}', A_{k-1}'', A_{k-1}''' (from fig. 6.13), for a given X_K (Ref. [7]).*

This optimisation problem can be solved by a standard, single variable, stochastic dynamic programming procedure [7]. It gives as a result the decision tables illustrated in section 6.3.2.1.

It is worth nothing that as a by-product, the water value is obtained by this procedure. It is defined as follows.

To any given couple of state variables at month k, X_k and $A_{(k-1)}$, an optimal value U_k is associated as stated before, and hence a profit (or cost) $F_k(U_k)$. The value dF_k/dU_k (computed on the cost-revenue function defined in A.7 of appendix 2) may be assumed as (incremental) water value ($/kWh), since an increase dU_k in hydro generation will of course entail an equal decrease in thermal generation or opportunity import (or an equal increase in opportunity export). Now, make an average of the various dF_k/dU_k corresponding to the various values of $A_{(k-1)}$, each weighted according to the probability of the $A_{(k-1)}$ values themselves (Fig. 6.14). This average is, by definition, the incremental water value $V_{(k,x)}$, which is a function of k and of the storage X_k (because U_k is a function of X_k, from decision tables). Since the terminal value function I has been assumed with decreasing derivative with respect to storage, it turns out that, for each k, $V_{(k,x)}$ is a decreasing function of the storage.

This parameter is considered by utilities A and B.

This definition of water value shows again the characteristic of a substitution value.

6.3.3.2 Hydro-thermal (hydro and thermal generation not less than 15% each)

Utilities D, E, F, G and H have been considered.

Utility D follows an operation policy type methodology (section 6.3.2.2).

The period considered is 53 weeks, and the elementary time interval is one week.

Three main steps may be considered:

1) definition of weekly cost curves $F_k(T_k)$, with $k = 1, \ldots , 53$. Since utility D has no opportunity markets (load or power exchanges, in the sense in which the opportunity markets were defined in A.3 of appendix 2), there is no profit concept, and the cost is associated only to fuel consumption; the control variable is the thermal generation T_k (MWh);
2) definition of constraints;
3) optimisation with constraints.

The definition of cost curves is done, according to the general concepts illustrated in A.7 of appendix 2, in the following manner, for the generic week k.

Assume that a forecast of load (chronological MW values for each hour, or each couple of hours) is available for week k (sequence of 168 or 84 values); the load curve is supposed to be net of power exchanges (utility D does not optimise directly this item of the balance); hence it must be covered by two items only (that is hydro and thermal generation).

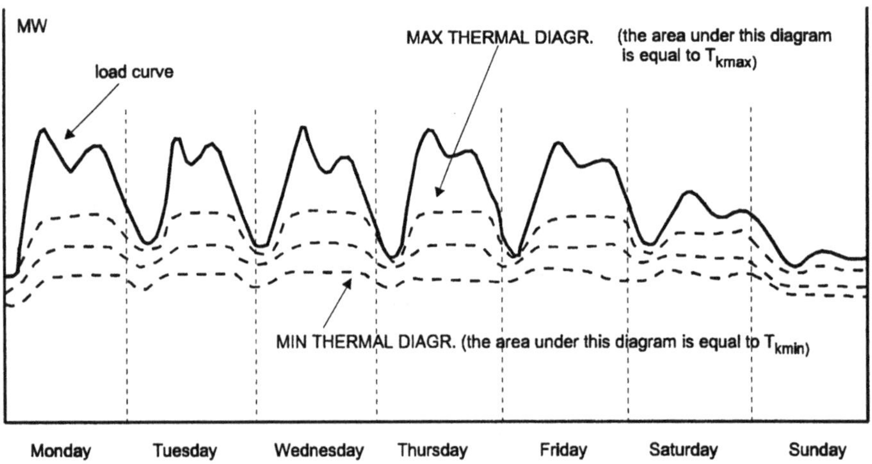

Figure 6.15: *Various thermal generation diagrams (Ref. [10]).*

Various thermal generation diagrams or curves are then considered (Fig. 6.15), of progressively increasing energy; the shape and number of such curves should comply with the following requirements:

1) at each hour the difference between load (average forecast minus 1 or 2 times the standard deviation) and thermal power shall not be higher than max hydro power available in case of wet hypothesis for hydraulicity;
2) at each hour the thermal power shall not be higher than maximum thermal power available (that is, not on scheduled maintenance);
3) each curve (to which corresponds a given value of T_k) shall be as flat as possible (hydro generation on peak-shaving duty);
4) the number of curves considered shall be sufficient to cover, in steps of a not too big value, the range from T_{kmin} (corresponding to low hypothesis for load and high hypothesis for hydraulicity) to T_{kmax} (max thermal availability).

Now perform the following steps:

1) choose one of the chronological thermal load curves defined above and shown in Fig. 6.15;
2) for each one of the thermal units not scheduled for maintenance make a chronological simulation, by Monte Carlo methods, of its availability (this simulation produces, on the basis of the transition rates characteristic of the unit, one on-off state sequence, all along the week, of the unit itself);
3) for each hourly interval make a choice (with an economic criterion) of a number of the units which were available in step 2, up to covering the total requested power; make a sharing of power between the chosen units, e.g. on the basis of equal incremental costs, and compute the total (weekly optimum) cost;
4) repeat the steps 2 and 3 an appropriate number of times (e.g. 100 or 500). Then for each unit not on scheduled maintenance, the average (over the 100 or 500 sequences sorted out) generation, consumption of fuel and cost may be computed, together with the corresponding probabilistic parameters (e.g., standard deviations). As far as the set of units is concerned, the average generation is of course T_k, and the corresponding average cost is $F_k(T_k)$;
5) repeat the preceding steps for all the other chronological thermal load curves (that is, with different values of T_k). As a final result, the (average) cost curve $F_k(T_k)$ is known (by discrete values of T_k), as well as its probability parameters (that is, for each value of T_k, average F_k and its standard deviation are known).

The reason for this procedure is to account for randomness in availability of thermal units, which has influence on the operation costs. As a matter of fact, suppose that some units are lower in size, lower in efficiency and of higher reliability, and that some others are of larger size and higher efficiency, and of lower reliability. Taking into account only the average availability of both small and large units, is likely to produce an underestimation of operation costs [8]. The method described above gives better results [9] (Fig. 6.16).

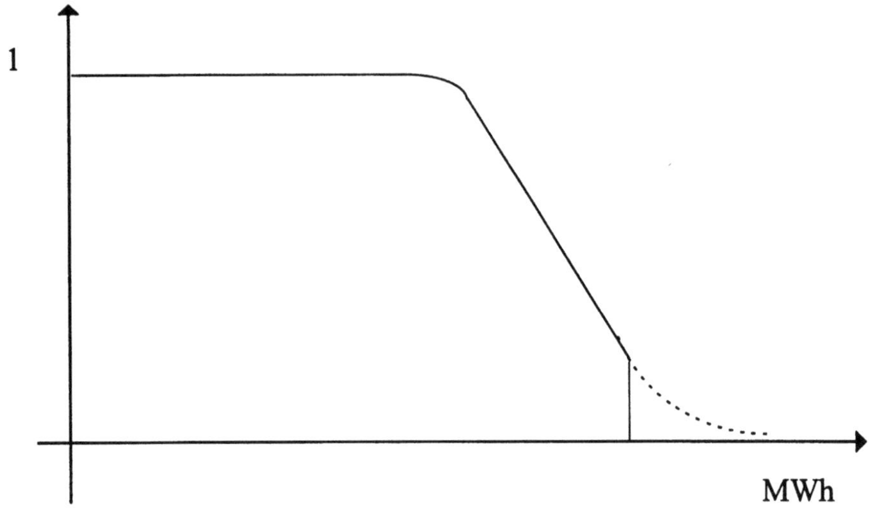

Figure 6.16: *Cumulated probability curve of generation for a thermal unit in the time interval k; the truncation (which may or may not be present) corresponds to maximum output (unit running at rated MW all along the time interval).Similar curves can be built-up for sets of thermal units (e.g. , those of the same fuel) and for sets of time intervals (covering for example one month, or the whole period under study) (Ref. [9]).*

Note that the load duration curves may not be used in this case, due to the key role of chronological simulations of step 2 above.

Note also that T_{kmax} will in general be less than that corresponding to full availability of thermal units, due to actual availability values of thermal units themselves taken into account in the Monte Carlo simulations.

The definition of $F_k(T_k)$ will of course be performed for each one of the weeks constituting the period under study.

The computing requirements (memory and time) are quite heavy.

The definition of constraints for utility D is illustrated in ref. [10] and [11]; the methodology consists in building up a cumulative, hydro plus thermal, probability curve of generation for the 53-week period, for the last 52 weeks, for the last 51, etc. These cumulative probability curves account for randomness of load, availability of thermal units and water inflows (simpler computing methods are used in this phase, with respect to those illustrated for the phase of definition of weekly cost curves). Having established before hand the accepted EUSE for each one of those time intervals (values given by management), a minimum permissible storage in the equivalent hydro reservoir at the end of each week is accordingly determined; this is a concept similar to that of alarm curve.

The optimisation is then performed, assuming a hypothesis (a forecast) for water inflows (weekly values) and for load. The optimisation procedure utilises the cost curves defined above, and is performed through a DP algorithm.

Because of the presence of an equivalent hydro reservoir, which can make transfers of energy from week to week (and month to month), the optimisation must cover the period of time (53 weeks) as a whole (that is, it cannot be decomposed in 53 separate and independent optimisations); the assumption of final storage equal to the initial one is generally assumed.

Note that, due to the deterministic assumptions of water inflows and the assumptions on given initial and final storages, the total amount of hydro generation (GWh) is known; because of the deterministic forecast of load, the total thermal generation is also known; the optimisation then consists in minimising the cost of given thermal generation (the presence of pumped hydro makes the problem a little more complex, but does not modify the concepts just illustrated).

One of the results is the schedule of thermal generation; it is worth stressing that, as illustrated before, fuel consumption (or associated cost) gets determined in average values and probability distribution; of course the average values depend also on the assumed values of hydraulicity. Another result is the weekly drawdown from the equivalent reservoir; etc.

The whole procedure is repeated every four or five weeks, always with a horizon of 53 weeks ahead.

Utility E follows a method quite similar to that illustrated for utility D; some additional simplifying assumptions are made on the representation of hydro generation, which to some extent is dealt with in a predetermined (heuristic) manner, the consideration of randomness being limited to load and availability of thermal units [12], [13].

The period considered is one year, and the elementary time interval is one month; the optimisation is split into 12 decoupled optimisations.

This simplification allows the representation of some power transfer limits between electrical sub-networks. The procedure consists in the use of Monte Carlo method to draw a number of possible system "states" (the value of load and available thermal units) in each time sub-interval of one hour (there is no sequential simulation, as in the case of utility D); for the optimisation of each state the algorithm used is linear programming, and the results (thermal generation per fuel type and area) are given in terms of average values and standard deviations.

The methodology followed by utility F is again of the type illustrated for utility D. The period considered is one year and the elementary time interval one week.

The use of network flow techniques allows a more detailed representation of hydro subsystem in the optimisation procedure, thus making possible a much more accurate accounting of hydro constraints (single reservoir storage limits, influencing etc.).

Randomness is accounted for availability of thermal units; water inflows and load are assumed at their forecasted values.

Weekly cost curves are not used; instead, a first attempt hydro schedule is tried, and correspondingly first attempt weekly thermal schedules and incremental costs are evaluated (with deterministic availability of thermal units). An iterative procedure between two steps is followed: modify the hydro schedule on the basis of thermal incremental costs, to levelise the incremental costs themselves; compute the new thermal schedules and incremental costs, until the convergence is reached. On this final (optimal) thermal schedule, a more accurate simulation of availability of thermal units is performed, to compute the expected energy generation of each thermal unit [14], [15].

Utility G is a hydro-thermal system with a dominant share of nuclear generation (pressurised water reactors). Two main steps are considered for medium-long term operation planning:

A. Six years horizon of study: definition of the maintenance and refuelling schedule for nuclear PWR units. The procedure used here is of the open-closed loop type, taking into account updated information on nuclear cores burn-up and maintenance and refuelling constraints. The particular characteristics of PWR must take into account the following:

- each reactor has to be shut down during six weeks for maintenance and refuelling, about once a year. This can be done when the nuclear fuel is between given limits of burn-up (anticipation and stretch-out);
- the average length of a cycle (including maintenance and refuelling) is by now about 14 months, and naturally increases when the unit is no longer base-loaded.

Thus, when a PWR is used as a cycling unit, in a load-following mode, or when it is on forced outage, the energy not used may help shift the next refuelling window. Special models have been developed to solve this problem on an adequate pluri-annual horizon of study [16].

B. 18 months to 24 months horizon of study: definition of the operation strategy of the seasonal hydraulic reservoirs and of the PWR units between two refuellings. The procedure used here is of the closed loop type, with two levels of study:

2) the aggregated, seasonal hydraulic reservoir, weekly poundage and run-of-the-river are first operated with SDP [17]. This first level gives all classical results of annual operation planning and moves the probability law of the marginal costs of operation to the second level;

3) the strategy for each national watershed is then optimised by operation feedbacks, with its detailed physical description, on the basis of the above mentioned marginal costs. By "feedback" is meant an operation rule which disaggregates the global withdrawal from the unique, equivalent reservoir into the partial withdrawals from the component watersheds. In particular, a "local feedback" is a rule allowing for example the determination of the storage of (or generation from) a particular reservoir: for a reservoir with water inflows coming from snow

melting, the more appropriate feedback could be from load trend (that is, generation from that reservoir will be increased or diminished in correspondence with positive or negative deviations of the load, respectively); for a reservoir with water inflows coming from rain, instead, the more appropriate feedback could be from storage itself (generation will be increased if the storage is increasing, and vice versa) [18].

The water values can thus be obtained for each seasonal reservoir and passed over to daily scheduling.

Utility H follows a methodology which consists in determining the water values according to a procedure similar to that described in section 6.3.3.1; account is taken of the randomness of the inflows and of the actual reliability of thermal units.

The starting point for the determination of the water values is at the end of the period under study; an arbitrary (to some extent) function $V_{(N+1,X)}$ is assigned ($/kWh as a function of storage X); then an optimisation is performed for the Nth time stage, assuming that incremental thermal generation cost is equal to incremental water value $dV_{(N+1,X)}/dX$, so that each trajectory from X_N to $X_{(N+1)}$ is supposed to be at constant incremental water value. Hence the function $V_{(N+1,X)}$ is projected back to give $V_{(N,X)}$ at the beginning of the Nth time stage (and end of the $(N-1)^{th}$ one). Then an optimisation is performed for the $(N-1)^{th}$ time stage, and so on, till time zero, that is, till determination of function $V_{(1,X)}$. If this last function is different from the starting function $V_{(N+1,X)}$, the assumption $V_{(N+1,X)} = V_{(1,X)}$ is made, and the whole procedure is repeated, until convergence is reached (assuming that $V_{(1,X)}$ must be equal to $V_{(N+1,X)}$ is a likelihood assumption, which is valid if the instants 1 and N+1 are for example distant by an integer number of years. In the case illustrated in sec. 6.3.3.1, instead, it was requested that the period under study be long enough to make it possible to assume $V_{(N+1,X)} = 0$).

In Appendix 2D an example of computation of water values according to this methodology is illustrated.

Once the water values have been so computed, the equivalent hydro reservoir plant may be considered as a thermal plant (with some appropriate constraints), and from this point onwards deterministic schedules may be produced, according to a methodology of the operation policy (section 6.3.2.2) [19].

6.3.3.3 Purely thermal (thermal generation higher than 85%)

The medium-long term operation planning procedures for purely thermal utilities are of course simpler than those for hydro-thermal ones. The key items are unit commitment algorithms, which should take into account the random availability of generating units, and allow for consideration of pumped storage plants if any.

Utilities L and M, both of which have pumped storage plants, use deterministic load forecast (chronological, in hourly steps), and consider a horizon of 1 or 5 years ahead. Pumped storage is run on daily or weekly cycles, for reserve or economy duties.

Sensitivity studies are performed, in particular with respect to rate of load increase.

Particular attention is paid to the preparation of merit order lists, which are used in unit commitment computations, and to the availability of various types of fuel (at costs varying in time). An important outcome of the procedure is the definition of fuel consumption of the various generating units, which is the basis for stipulating medium-long term fuel purchase contracts.

A description of the operational planning procedure of utility M is now given.

The medium term plant schedules are computed for each year of the five-year period ahead.

A. A suite of programs perform the computations corresponding to two main functions. These are to calculate:
   (a) the energy production and heat demands for each power station;
   (b) a minimum transportation solution to satisfy station heat demands determined in (a) from available fuels.

B. The programs perform a number of iterations. The marginal heat costs determined in A(b) are used to determine a merit order which is used at each successive iteration for scheduling plant in A(a). Up to 10 iterations may be necessary to determine the lowest cost solution.

C. Plant scheduling method. The scheduling method used in A(a) is based on the same program used in the short-term day to day control room phase. The only significant difference is that, instead of running the program for the next few days ahead, it is run to cover a period of one year based on 12 monthly simulations. Some features are as follows:
   - Demand. For each year the power and energy demand for 365 days are modelled in full. Each day is broken into 12 two hour steps. The daily demand curves are obtained from a historical data bank with some allowance for weather sensitivity as required.
   - Generation. Some 80 power stations are modelled; these are predominantly coal fired, but also include some nuclear, oil, pumped storage and gas turbines stations. Average forecasts of availability of generating plants are used; these reflect the plans for closing down old plants, commissioning new plants, seasonal cold plant regimes, maintenance outages, and an allowance for plant breakdown. A deterministic approach is used and random availability changes are not modelled due to complexity of trying to attempt this with the size of the simulation involved.
   - Generation plant characteristics. The full range of parameters for generating plants are modelled as appropriate, i.e. plant output capability, run-up and run-down rates, minimum on and off times,

synchronising and maximum generator level, time between sets on and off, starting-up costs, plant flexibility, dual firing, stabilising oil burn and fixed and incremental heat rates.

- Trading. Long-term trading arrangements are modelled and various opportunity options are investigated.
- Transmission system. The main transmission system is simulated and usually divided into eight key zones to model power constraints associated with maintenance outages. The number of transmission zones may be increased if required. Typical transmission loss factors are modelled for all generating plants.
- System reserves. A predetermined system reserve policy is modelled which varies with the season of the year. This is generally based on a spinning reserve comprising pumped storage plant during the day time and steam plant at night. This is backed up with gas turbine plants in a standby mode.
- Merit order. For the first iteration of the simulation run a previous merit order is used to start off the process. For the second iteration the heat cost is input to the merit order used by the scheduling program from the fuel allocation program. Two types of merit order are in use, these being based on the system per unit cost multiplied by the incremental or full load heat rate. The full load cost is used in the scheduling algorithm to bring plant on and the incremental cost is used to allocate output to plant brought on. Up to three incremental steps may be modelled for various coal-oil fuel mixes if required.
- Scheduling method. The scheduling program initially selects plant with the lowest merit order cost based on the full load heat rate to meet peak or trough demands. For a selected amount of marginal plant these costs are then adjusted by various penalties: (a) for starting-up or shutting-down plant; (b) other penalty costs associated with the characteristics of plant with parameters that do not meet ideal system requirements. The program then modifies the selection of plant based on these adjusted costs and allocates output based on incremental costs.
- Pumped storage plant. The operation of pumped storage plant is simulated in the scheduling process as follows: (a) the scheduling procedure described above is carried out; (b) the marginal costs are passed on to the pumped storage routine; (c) the procedure described above is repeated with the load curve modified by pumped storage plant operation.

D. Fuel allocation method. The input to fuel allocation program is as follows:

- Input from the scheduling program. The annual on-load and off-load heat required at each station is input into the fuel allocation program from the scheduling program;
- Fuel and transport data. Fuel for each fuel source availability, calorific value and pit head costs are modelled together with the transport routes

available from source to power station identifying the mode available and the cost of each mode. Up to 200 fuel sources are modelled together with some 1600 delivery routes. Other parameters modelled include the cost of fuel and ash handling at each station, fuel quality and capability constraints of each transport mode and route together with the requirements of fuel stock level. The stock levels may be set to achieve some strategic aim.

E. Typically the type of simulation described above is usually carried out for the current year and one year ahead. For the years beyond this a simpler process is used based on a fixed merit order adjusted to reflect future fuel agreements. A range of sensitivity studies are carried out to cover uncertainties.

F. Associated operational planning activities. From the results of the energy planning simulation much useful economic information can be derived using special analysis tools. This includes assessment of monthly weekday-weekend and annual system marginal replacement costs for each generating unit, the associated load factors and plant utilisation. This information is then input into other operational planning activities which include generation and transmission maintenance schedules, management of plant capacity for seasonal cold plant, operating regimes and plant refurbishment, trading arrangements, bulk supply tariff work, and national spare plant strategies. Another important activity is the assessment of plant margins.

G. Plant margins assessment. In addition to energy management studies, plant margin assessments are carried out to ensure that the amount of plant made available for each week or month of the year meets the operational standards of security. These balances are simple checks and may be performed relatively quickly.

H. For the current year and one year ahead plant margin balances are carried out every two months for each week of the period. The first six weeks of the current year are reviewed on a weekly basis as part of shorter term work.

6.3.4 Conclusions

The methodologies followed in medium-long term operation planning (1 to 5 years ahead) may be grouped in three broad categories: operation strategies, operation policies, deterministic optimisations.

For reasons of practicality, reflecting in some cases the organisation of the utility, the methodologies may frequently consist in a mix of the above three categories.

The first one seems to be preferably followed by hydro dominant utilities; it gives operating guide lines for the next elementary time interval (in most cases, next month) through decision tables. In principle the decision tables (one for each elementary time interval or month) do not need to be updated if the random parameters (water inflows, load, availability of thermal units) keep within the historical limits (closed loop characteristics). The main theoretical problems are those referring to the representation of the hydro system, which must be more or less aggregated. It is often used to account for randomness of water inflows only (load and availability of thermal units are considered deterministic).

The second category leads to the determination of generation schedules covering the whole period under study. Randomness is taken into account in a ampler way than in the first category, normally in the phase of definition of appropriate constraints; the last are then introduced in the procedure for determination of schedules (this last procedure is substantially of a deterministic nature, hence it is to be expected that optimality is approached with less accuracy than with the first category of methods). The above-mentioned constraints have the main object of ensuring an appropriate degree of security. Hydro-thermal utilities show a preference for this methodology or, in some cases, for a mix of the two. The results must in general be updated (for example, every month), taking into account major departures of parameters from forecasted values (open-closed loop characteristics).

The third category is fully deterministic. Randomness is in some way accounted for by running the procedure with some alternative values of random parameters, such as load increment rate, that is, performing sensitivity studies (open loop characteristics). Purely thermal utilities often use this kind of methodology. It should be stressed that for such utilities the overall economy of operation is greatly affected by fuel costs and currency exchange rates (if fuel is imported).

References

1. CIGRE SC39-WG03 *Operational planning functions - Short term scheduling: present practices and trends* - Electra CIGRE, May 1986.
2. H. Duran, C. Puech, J. Diaz, G. Sanchez - *Long term generation scheduling of hydro thermal systems with stochastic inflows* - IFAC, Rio de Janeiro, 1985.
3. Expansion planning for electric generating systems: a guidebook - *I.A.E.A. technical reports series*, Vienna, 1984.
4. A. Turgeon - *Optimal operation of multi-reservoir power system with stochastic inflows* - Water resources research, April 1980.
5. M. Pereira, L. Pinto - *Stochastic optimisation of a multi-reservoir hydroelectric system: a decomposition approach* - Water resources research, June 1985.
6. P. Lederer, Ph. Torrion, J.P. Bouttes - Overall control of an electricity supply and demand system: a global feedback for the French system - *11th IFIP Conference*, Copenhagen, 1983.
7. R. Pronovost, J. Boulva - Long-range operation planning of a hydro-thermal system: modelling and optimisation - *Journal of Canadian Electrical Association*, 1978.
8. R. Fancher, T. Guardino - *Probabilistic production costing with load-shifting*

resources using discrete approximations - IEEE trans. on PAS, August 1984.

9. G. Fusco, L. Vergelli - *Monthly generation forecast for thermal units of given reliability* - PSCC V, Cambridge 1975.
10. A. Di Perna, E. Mariani - *Medium term production schedules in a hydro-thermal electric power system* -PSCC V, Cambridge 1975.
11. Outline of a procedure for the determination of expected unserved energy in a hydro-thermal electric power system - *Internal CIGRE SC39 - WG03 paper*, August 1983.
12. M. Huggins, M. Mirsky - *Optimal energy transfers in interconnected electric systems - IEEE transactions on Power Apparatus and Systems*, November 1985.
13. L. Wang, K. Gallyas, D.T. Tsai - *Reliability assessment in operational planning for large hydro-thermal generation systems* - IEEE trans. on PAS, December 1985.
14. D. Sjelvgren, S. Andersson, T.S. Dillon - *Optimal operations planning in a large hydro-thermal power system* - IEEE trans. on PAS, November 1983.
15. S. Andersson, D. Sjelvgren - *A probabilistic production costing methodology for seasonal operations planning of a large hydro and thermal power system* - IEEE trans. on Power Systems, November 1986.
16. A. Merlin, P. Roussel - *Optimisation of the refuelling schedule of a PWR electric power generation system: the PLANUM model* - PSCC July 1981.
17. V. Garabedian, F. Meslier - *The GRETA model: a decision making tool for operation and planning of generation systems at EDF* - EPES July 1979.
18. P. Colleter, P. Lederer - *Optimal operation feedbacks for the french hydropower system* - CORS.TIM.ORSA May 1981.
19. F. Alonso et al. - Yearly operation planning of a hydro-thermal system with regulating reservoir - *Internal report of department of energy management*, Iberduero S.A., 1984.

CHAPTER 7

LOAD MANAGEMENT AND METHODS OF MEETING PEAK DEMAND

7.1 General

In many developing systems, the concept of reserve has not taken a concrete shape with the result that not only are no reserves maintained or even efforts made to build up reserves over a period of time, but also more load is taken on the system than the matching available generation thus compromising on the standard system frequency. This tendency is detrimental to satisfactory system operation and is injurious to power plant equipment and consumers' installations alike. The ill-effects of under-frequency operation are dealt with in Chapter 7 of [4] in some detail.

Continuous matching of the load demand with the generation available is load management. Any mismatching leads to variation in system frequency. Measures have to be taken to maximise the power availability and minimise the system demand so that the mismatch between the demand and the availability is reduced and the system frequency is within acceptable limits. Sustained efforts to reduce the down-time for maintenance and forced outages as well as partial capacity reductions would increase the net generation capacity available. A parallel exercise has to go on continuously to keep the load demand within the generation available.

In developing systems, in view of the serious constraint on resources, the growth in the generation capacity is unlikely to overtake the load growth in the foreseeable future. The power demand 'suit' has therefore to be cut according to the 'cloth'. Even utilities in the developed countries with surplus power available are increasingly adopting the technique of load management with a view to conserving resources.

7.2 Flattening of Load curve

Load management is a means to derive economic advantages in the operation of power plants. For this purpose, efforts are made to obtain as flat a load curve as possible. Flattening the load curve leads to;

- avoidance of start-up and shut-down of power plants.
- less requirement of regulating capacity
- reduction in transmission costs
- avoidance of overload of equipment in parts of the system.

Fig 7.1 shows the lines of load management.

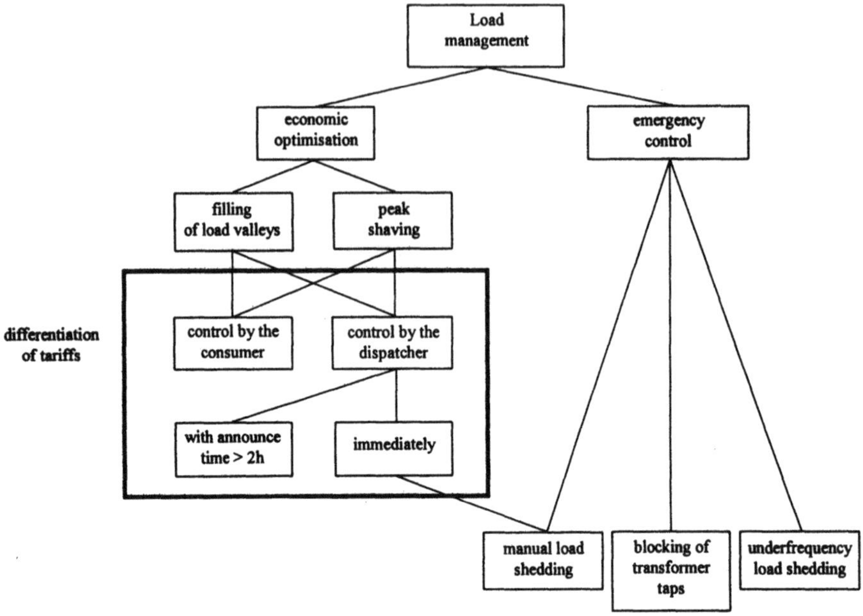

Figure 7.1: *Objectives of load management*

The flattening of load curve may be achieved :

- partly by differentiation of tariffs so as to encourage the consumers to use less energy during the system peak hours, and
- partly by giving an economic advantage to the consumers through load reduction by the dispatcher.

7.2.1 Differentiation of tariffs

In most utilities there is differentiation of tariffs depending on the time of the day. Normally, the tariff for energy (and in some cases, power especially for the industrial consumers) is lower in the off-peak hours. The off-peak hours are defined with reference to the type of day (working day, Saturday, Sunday / holiday) and in some cases seasonal considerations. This differentiation in tariffs leads to control of the load by the consumer without any action of the dispatcher.

The differentiation of tariffs has an influence both on the filling of the load valleys and reduction of the peak demand. A typical result is shown in Fig. 7.2 for three industrial consumers, where the tariff is higher between 7.30 hrs and 17.30 hrs.

A survey made by WG39.03 of the CIGRE has shown that some utilities have realised an estimated reduction in daily peak up to as much as 15%.

The survey has also shown that off-peak heating is resorted to in a number of utilities. Special contracts are entered into without a demand charge and with an energy charge lower than normal. Fig. 7.3 shows the results achieved by one utility through this method.

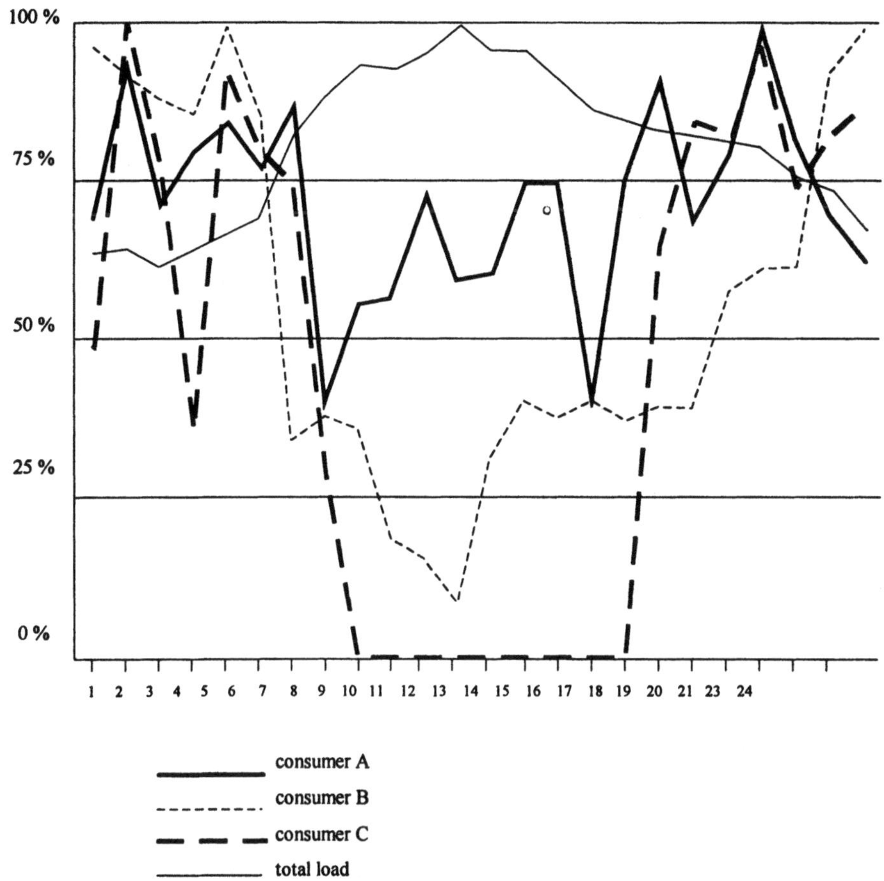

Figure 7.2: *Load curves of industrial consumers influenced only by differentiation of tariffs.*

For control of off-peak heating, a large number of consumers (about 12,000 in the case of the utility covered in Fig. 7.3) must be switched in. There are two possible ways of doing this : either by clock switches installed at the consumer site or by ripple control.

7.2.2 Ripple Control

The ripple control system is used in some systems to flatten the daily load curve and is particularly attractive in urban concentrations where a large number of residential, commercial and industrial loads can be controlled. A high-frequency ripple or alternatively a kink introduced into the sine wave is used to send a coded message to actuate relays installed at non-essential loads of consumers such as residential water heaters and air-conditioning, some commercial air-conditioning, space-heating, etc., which are automatically switched off. The system does not cause any disturbance to daily living as would be the case during load shedding through under-frequency relays during emergencies.

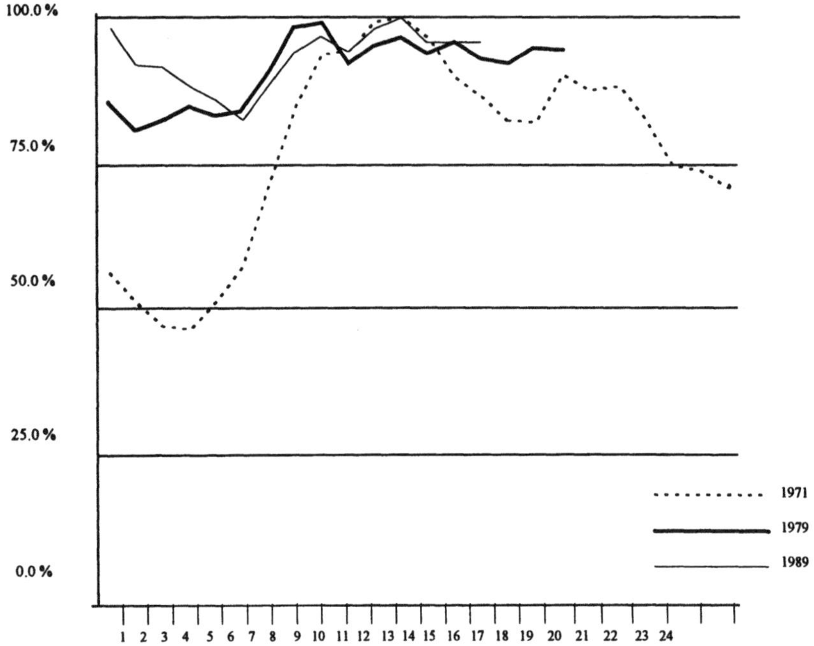

Figure 7.3: *Filling of load valleys by off-peak heating*

7.2.3 Under - Voltage Operation

A practice adopted in some power systems to reduce the load during peak load time is to reduce the voltage at the consumers' premises. In this scheme, the voltage on the low voltage side on the step-down transformers at selected sub-stations is purposely reduced by lowering the taps. Different types of loads respond differently to voltage variation. Filament lamps will have approximately a 1.6 % wattage reduction for every one percent drop in voltage and the fluorescent lamps approximately a 1.4% wattage reduction. Resistance loads absorb about 2% less power for one percent reduction in voltage. In the case of induction motors there are three types: i) constant power (mixers), ii) constant torque (compressors, refrigerators, air-compressors, etc.) and iii) variable torque (blowers, fans, centrifugal pumps, etc.). The power reduction in these motors varies from 0.4 to 1.45% for a 2.5% voltage reduction from normal voltage. It has been observed in the case of Canadian and American power systems that 1% voltage reduction results in a reduction of 0.5 to 1.1% of system demand.

7.3 Characteristics of Loads

The loads generally fall under the following categories :

- Three- shift industries and continuous process industries present almost a flat load demand.
- Two-shift and single-shift loads require supply during certain hours.
- Lighting load requires supply during the evening and night hours and a small percentage of it during the day in offices and commercial establishments.
- Seasonal loads like agricultural, cotton ginning and pressing mills, oil mills, etc., require supply only during certain seasons.

7.4 Measures for Load Management

The following measures are adopted for load management in some of the developing systems:

1) Staggering of weekly holidays and lunch recess.
2) Deferring of loads - Agricultural consumers to avail of power supply during certain hours. For example, in Tamil Nadu, irrigation pump sets are grouped as follows :

   Group I : From 6 AM to 12 Noon

   Group II : From 12 Noon to 6 PM

   Group III : From 10 PM to 6 AM next morning

   Power is not supplied to any group between 6 PM and 10 PM., that is, during the evening peak hours. The groups are rotated once in 10 days, so that no single group is given supply only in the night during a month.
3) Single- or two-shift industries not to consume power during peak hours.
4) Restrictions on air-conditioning during peak hours.
5) Continuous industries to have a cut in their demand (and also energy, if required).

7.5 State-of -the-art in the USA

In the USA, load management is receiving considerable attention because of the extraordinary high costs of installing new generating capacity and willingness of the consumers to conserve.

The following criteria have been spelt out for load management programmes :

1) It must be able to reduce demand during system critical load periods.
2) It must result in a reduction in net generation requirements, purchased power and / or fuel costs.
3) It must have an acceptable cost / benefit ratio.
4) Its operation must be compatible with system design and operation.

5) It must operate at an acceptable reliability level.
6) It must have an acceptable level of customer inconvenience.
7) It must provide a benefit to the customer in the form of reduced rates or other incentives.

With proper design, load management is expected to provide a portion of the operating reserves, thus reducing the spinning reserve requirements. The overall system reliability, it is stated, can also be materially improved by load management.

7.6 Swedish Project to Optimise Energy Consumption Pattern

The Swedish Kraftnat has launched a major project designed to chart home and industrial energy consumption in Sweden and study various means of reducing and optimising electricity use. The study was prompted by prognosis indicating that the price of electricity will rise by the mid-1990's, partly as a result of the planned phasing out of nuclear power in Sweden by the year 2010. The project, named "Mission 2000", will explore the possibility of holding down electricity costs by rationalising energy consumption.

The project is being carried out in two phases. During the first phase which will take two years, 2000 property owners in three northern and central Swedish municipalities will be offered, free of charge, inspection of the efficiency of their energy consumption. Another 4000 households in the same municipalities will be interviewed with regard to their energy use habits. In order to study industrial energy consumption, all small and medium-size industries in Vargarda in south-western Sweden will be invited to take part in the project.

The second phase of the project will utilise the background material collected during these first two years to plan a five-year, full-scale energy rationalisation project in one of the three municipalities. Simultaneously, the measures developed to optimise industrial energy consumption will be generalised to other areas of the country. The most important factor within industry is to make industrial processes more efficient.

7.7 Methods of meeting peak demand

Most modern power systems are a mixture of conventional hydro, thermal and nuclear plants. The problem of supplying peak load in such systems is becoming increasingly important. It is really the advent of nuclear power that has caused a multiplication of system planning complexities, one of which is the problem of meeting peak load. Conventional thermal units have a considerable degree of flexibility on the whole, even allowing for their long start-up time and their limitation with respect to rate of loading. But a nuclear plant is constrained by the nuclear physics of reactor design to operate at a fixed level although in recent years, the nuclear plants also perform daily cycling (load follow), for example, at Electricité de France. This calls for the remaining conventional thermal and hydro plants to cater for all variations in load demand as well as the system peak load.

Thus, run-of-the-river, nuclear and high-merit thermal plants are base loaded, the reservoir type of hydro stations being used for peaking purposes, the intermediate portion of the load curve being filled up by low-merit thermal stations and poundage type hydro stations (Fig.7.4).

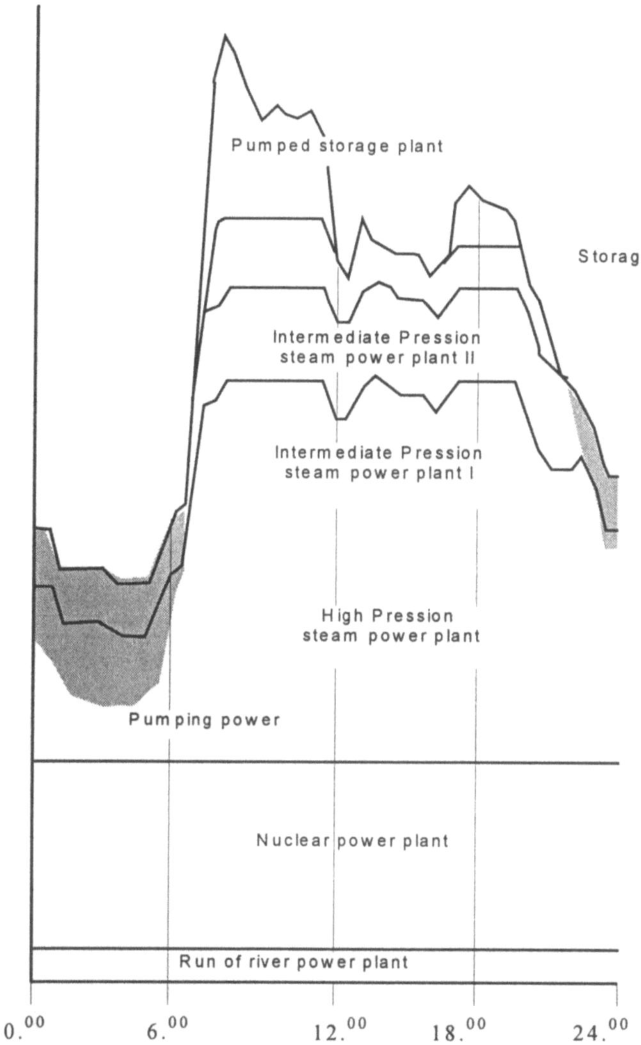

Figure 7.4: *Typical load curve for a working day with pumped-storage operation*

The evolution of the shape of the load curve on a typical winter day over a period of 15 years in Great Britain shown in Fig. 7.5 illustrates how the peaks have been contained through various measures.

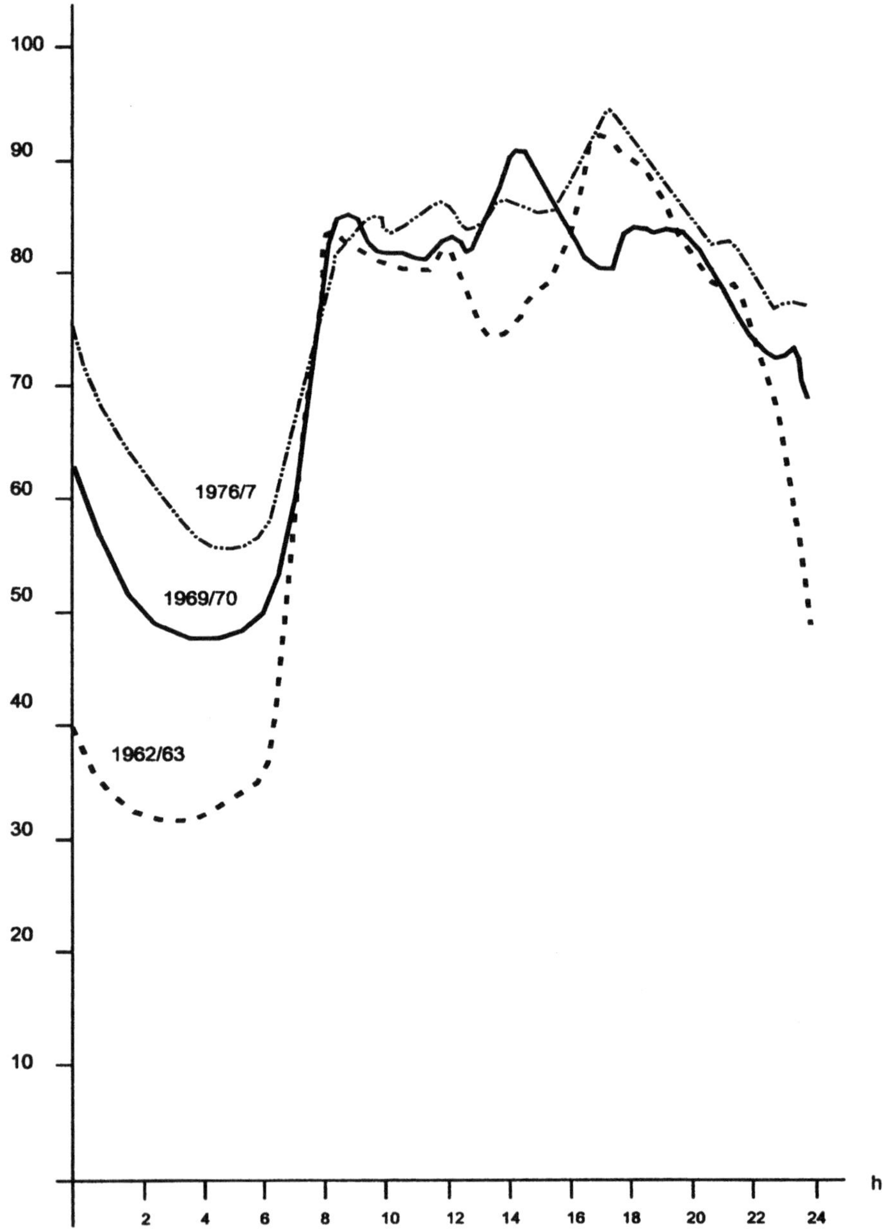

Figure 7.5: *Typical winter days expressed as % of annual average cold spell peak demand*

In the British system the minimum demand (between 23 and 5 hours) has improved significantly from 31% to 55% during the period 1962-63 to 1966-67. Further, at 24 hours while the demand was about 50% of the cold spell peak demand in 1962-63, it increased to about 79% in 1976-77.

7.8 Pumped Storage Plant

7.8.1 Advantages and Storage Cycles

In systems where pumped-storage hydro plant is available, they can be conveniently used for peaking purposes.

So important is this peaking power that some future hydro-electric sites may be developed not as conventional hydro plants, but primarily to supply peaking power, that is to say, as storage or pumped-storage installations. As a simple example, consider a hydro site with limited storage where the flows justify a conventional hydro installation of 100 MW. With peaking duty in view this plant may in fact be designed for an installed capacity of 400 MW for operation during the four hours of peak load only, storing water during the remaining twenty hours. Again, if some lower reservoir is available it may be possible to install some pumping capacity at the station, to justify an installed capacity of 600 MW or more for peaking purposes. This shift in philosophy with respect to future system planning means that power plants or units will not necessarily be expected to be both energy and power producers, they will generally be either one or the other. Among peak load plants, pumped storage stations occupy a special position, since they are able to reverse their role and take load during off-peak periods.

Pumped storage scheme, in its simplest form, consists of an upper reservoir and a lower reservoir with facilities for pumping water from the lower to the upper reservoir during off-peak hours and for discharging the water from the upper reservoir through a turbine to the lower reservoir during peak hours. Various combinations of pump, turbine, generator and motor are employed. The capacity of the pumped storage plant which can be accommodated in a system economically and technically depends on the characteristics of the demand and of the other generating plant. To minimise capital costs the same electrical machine is generally used as the motor for pumping and as the generator, so that the capacity for pumping and for generating must be similar. The overall efficiency of 70% or more which can be obtained with modern pumped storage plant, means that the production cost per kWh is 1.43 times the pumping cost per kWh. This also means that the pumped storage plant should not replace other generating plant with a production cost lower than 1.43 times the pumping cost.

Pumped storage plant has a number of advantages to which it is difficult to give a monetary value, but which nevertheless increase security and operating convenience. They are :

1) The plant has the same characteristics as conventional hydro-electric plant - quick starting and loading, but is independent of seasonal variations in rainfall.
2) It provides load for base load plant which may be unused during the period from midnight to 07.00 hours.

3) The pumping load is rejectable and is a substitute for spinning reserve capacity. Where the generating units are large as a percentage of the maximum demand of a system it is difficult to provide effective spinning reserve during the light load periods. The pumping load has the double advantage of increasing the minimum demand and providing rejectable load.
4) It provides an emergency stand-by supply for starting up steam plant after a complete system shut-down.
5) If it is located near the load centre it can improve the utilisation (load factor) of the transmission system and can be used for voltage control.
6) The addition of pumping to a conventional storage type hydro-electric installation can increase the load factor or justify increased MW capacity.

The way in which a pumped storage scheme is operated depends on the capacity of the upper storage reservoir. Since peak and off-peak thermal costs vary throughout the year, the larger the storage, the greater will be the opportunity to displace the most costly thermal plant with the cheapest pumping energy. Three distinct storage cycles are possible depending on the degree of storage available, namely, daily, weekly and seasonal.

7.8.1.1 Daily Storage Cycle

Because of the high cost of storage many of the existing installations have storage designed to permit operation on a daily cycle. This requires pumping overnight to provide generation for the following day. The total available upper storage is usually equivalent to about 4 to 6 hours of full load generation of which 1 to 2 hours of storage is usually held in reserve against emergencies; the reminder is normally fully utilised over the 24 hr. period.

7.8.1.2 Weekly Storage Cycle

Where the cost of additional storage is small it may be economical to provide sufficient storage to allow a weekly storage cycle. This is done for example in the Scottish scheme at Cruachan. The value of this cycle lies in the ability to pump at high load factor during the weekend with plant which is surplus only at that time. The water thus stored during the weekend is released throughout the following weekdays and is only partly replaced by a limited amount of weeknight pumping. The storage provided at Cruachan is equivalent to about 20 hrs. of full load generation and a small proportion of this is always held in reserve.

7.8.1.3 Seasonal Storage Cycle

The third operating cycle using seasonal storage is theoretically the ideal but rarely obtainable in practice owing to the cost of sufficient storage. The cost of incremental storage in excess of that required for a daily cycle has to be balanced by the reduction in fuel costs. The Schluschsee scheme in Germany has seasonal storage which allows summer hydro spill generation from snow melt to provide a large proportion of the pumping energy.

7.8.2 Types of pumped storage equipment

There are two types of equipment arrangement available for pumped storage development. One is the separate pumps and turbines coupled to a common motor / generator and the other is reversible pump-turbine units.

In most respects reversible generator / motors are very similar both in design and construction to generators for conventional hydro-electric generating stations. Special features are however introduced to cover the following requirements :

a) As a single runner serves for both turbine and pump, the unit must be started, run up to speed, and synchronised with the system by electrical means.

b) The generators / motors must operate equally satisfactorily in both directions of rotation; they must be capable of withstanding frequent load cycling.

An idea of some of the major existing pumped storage developments in some countries of the world is given below :

Sl. No.	Country	Station	No. and type of sets	Generating head m	Station Capacity (MW)	Year of commissioning
1.	Luxembourg	Vianden	9FS	278	900	1964
			1RT		196	1971
2.	U.S.A	Muddy Run	8RT	125	800	1967
3.	Spain	Villarino	4RT	382	540	1969
4.	Japan	Azumi	4RT	135	436	1969
5.	U.K.	Cruachan	4RT	362	400	1966
6.	Germany	Sackingen	4FS	400	360	1967
7.	Italy	Presenzano	4RT	480	1000	1990
8.	India	Kadamparai	4RT	374	400	1989

F = Francis turbine, S = Separate pumps, RT = Reversible turbine.

The largest existing pumped storage development is the Vianden station in Luxembourg with a station capacity of about 1100 MW. This scheme is an outstanding example of international co-operation. The Sociétè Electrique de l'Our, the Luxembourg company formed to construct and operate the pumped-storage installation, is financed by Luxembourg, German, Swiss, French and Belgian interests. One of the main shareholders is RWE of Essen, the German power company to whose network Vianden is connected.

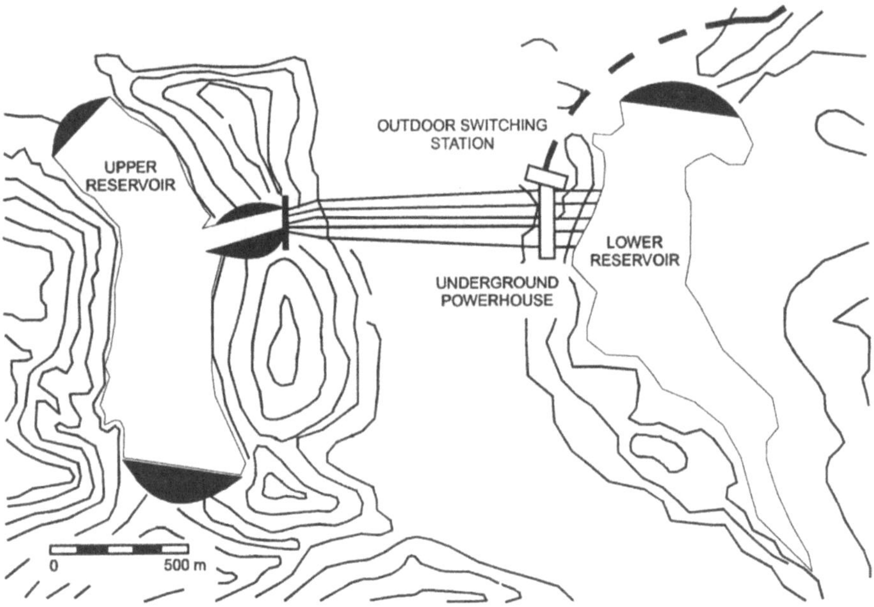

Figure 7.6 (a): *Pumped-storage plant project for 1000/2000 MW*

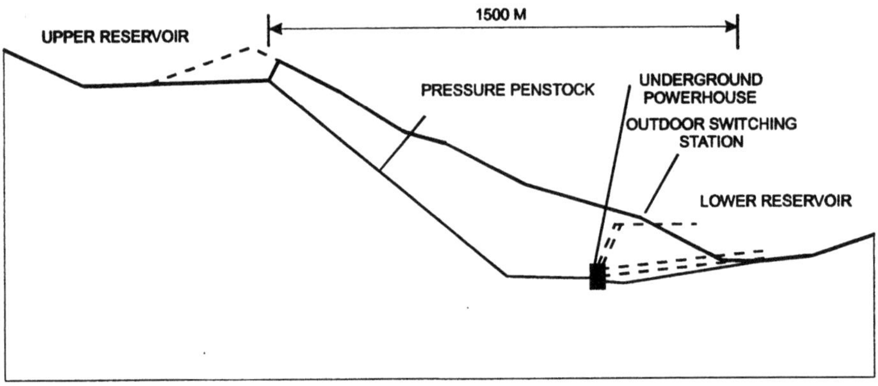

Figure 7.6 (b): *Pumped-storage plant project of fig. 7.6(a): vertical section.*

Vianden is primarily a peak-load and frequency regulating plant with a daily cycle, and is capable of providing a very rapid load swing of 1,520 MW, from a consumption of 620 MW when all the pumps are operating, to an output of 900 MW during turbine operation. The input of the plant is derived from coal or lignite-burning thermal stations of the RWE system in north-west Germany and its output is returned to the German grid, which has local interconnections with the Dutch, Belgian, French and Swiss networks. The plant is controlled from the RWE system-control centre at Brauweiler, near Cologne.

Of particular interest is a pumped-storage project in Germany in which two valleys are used which have a difference in elevation of 450 m and a horizontal distance apart of 1500 m, with a capacity of 1000 MW / 2000 MW for eight operating hours and 8 GWh of energy (Fig.7.6).

References

1. Jaeger, C. " Peak Load and pumped storage power stations", *Siemens Review*, XXXVI (1969) No. 8.
2. *Load Management - How will operators want to use it ?* IEEE Transactions Vol. PAS - 102, No. 6, June 1983.
3. WG39.03 *Load Management*, ELECTRA CIGRE No. 145, December 1992.
4. E. Mariani and S.S. Murthy, *Control of Modern Integrated Power Systems.* Springer-Verlag, London, 1997

CHAPTER 8

SECURITY AND RELIABILITY OF ENERGY CONTROL SYSTEMS

8.1 Introduction

Energy Control Systems (ECS), also called in French Systèmes de Conduite de Réseaux Electriques (SCORE), are the terms used by UNIPEDE, in developing the criteria related to the improvement of security and reliability of ECS or SCORE.

The purpose of the Energy Control System is to ensure the quality and security of operation of the power system covering a defined geographical area. The performance of ECS influences the quality, security and economy of the power system. This presupposes that the power system which the ECS is intended to control has been planned according to a well-defined philosophy and lends itself to reliable and optimum operation.

Following the historic Northeast Power Failure on November 9, 1965, the Federal Power Commission of U.S.A. made several important recommendations for preventing major power failures and one of the recommendations was to set up strong regional organisations for co-ordinating the activities of individual bulk power supply systems from planning through operation. Accordingly, nine Regional Reliability Councils and the North American Electric Reliability Council (NERC) were formed which were assigned this responsibility. The Operating principles laid down by the NERC, which are of general interest to operational planning and control engineers, are given in Appendix 2 of [3].

A secure and reliable ECS implies that the system is continuously available, providing information for decision-making and for executive instructions even under extreme operating conditions and within pre-determined response times. It is assumed that a computerised information system, in real time operation and with some automation in the dispatching functions, is available.

8.2 Organisation

8.2.1 Structure of the ECS

ECSs are generally organised following a tree structure with a hierarchy of several levels, the lowest levels being nearest to the network (substations and generating plants), and the highest level represented by the Central Control Centre co-ordinating generation, transmission of the bulk power system and interconnection with neighbouring systems.

The following alternative organisations can be identified in the ECS :

- One centralised with all monitoring and control functions carried out in the Central Control Centre ;
- One decentralised in which some of these functions are assigned to regional and eventually local control centres. Two alternatives are possible in this category, viz.,
  * one with centralised computers,
  * another with decentralised computers.

Accordingly, the following three organisations are possible and the impact of these organisations on the security and reliability of the ECS should be considered:

(a) Centralisation of functions and centralisation of facilities.

(b) Decentralisation of functions and centralisation of facilities.

(c) Decentralisation of functions and decentralisation of facilities.

In general, centralisation leads to a better use of equipment, that is, to better compatibility and quality; on the other hand, decentralisation leads to a better security and safety because it distributes more evenly the work and the responsibilities between the different centres. Hence, the worst accident, like the complete loss of activity of a centre, has less impact on the whole and may be covered more easily through back-up procedures.

The way of distributing the real time tasks, and their back-up, is obviously of major importance for the security of the ECS. It also has consequences on the manner of preparing for organised degradation, when one accepts the loss of economy before quality, quality before security, and security before safety.

The progressive computerisation of the whole system will substantially enlarge the range of possible solutions to the distribution of tasks. However, when a task is split among two or more levels, the higher levels should be devoted to co-ordination so that the loss of one part does not impair the main functions of lower levels.

8.2.2 Role of man

The control centre's operators or dispatchers form a constituent part of the ECS. Not only does the man have an influence on the functioning and on the interpretation of the information supplied by the system, but he is also influenced in his action by the possibilities and the limits of the system. So, the security and reliability of the ECS will be influenced by the man's capability while the man's ability to operate the system in a reliable and secure manner will be influenced and limited by the characteristics of the system.

The operator needs to be trained for operating the system in the best possible manner, for retrieving easily the information required to solve the problem he is facing and for making use of this information for decision-making. He is an important factor in sustaining the security of the ECS when some component or processing fails and has the responsibility for handling the failure and raising the level of reliability by using emergency procedures.

However, the ability of the man to increase the security of the system can be restricted for two reasons. One, the ECS is so reliable that the high level of security he is used to depending upon this system itself makes him lose his familiarity with the emergency procedures he has been trained for use in case of failure of the ECS; two, the structure of the power system is becoming more and more reliable with less and less incidents so that he is less prepared to face an incident on the power system whatever be the ECS facilities.

Because of these reasons periodical refresher training courses are required for the operators. Simulation of incidents executed either by an off-line simulator or by using spare capacity on the computing and display facilities would be an useful aid to keeping the operators up to date.

The security of the ECS must also be assessed periodically. It is controlled by men and they could make mistakes. Tests and checking of the correctness and precision of the operator's action could be effected by the ECS equipment.

For the other categories of personnel too, many of the criteria defined for operators are appropriate. Also, close co-operation and co-ordination is necessary together with regulations and procedures to ensure the content and timing of reports of abnormal situations amongst the different personnel involved. For example, the computer's operators will execute program running and software maintenance according to the dispatcher's requirements. The telecontrol and telecommunication maintenance staff will receive all information available to the dispatchers and connected with their activities while informing the dispatchers of all abnormal situations concerning their system.

8.3 Administration

8.3.1 Formulation of Orders and Instructions

The orders and instructions concerning operation may be categorised as : man to man, man to machine, and machine to machine. Three sub-systems are involved in this process:

1) The Operational or decision sub-system
2) The Information sub-system.
3) The Executive sub-system.

8.3.1.1 The Operational or decision sub-system

The different levels of operation and the departments using information entering the ECS or provided by the ECS will determine the information flows and the connection between the various parties. A procedure has then to be found which generates and transmits instructions comprehensively, clearly and briefly so that any error in the formulation and interpretation can be avoided. Time delays must be minimised in order to avoid the risk of further dangerous situations developing in the meantime. To this end it is advisable to formulate an "operation code" establishing in advance the sequence of various instructions and orders as demanded by the various situations. These orders will be learnt by the operators or coded in the computers.

Using such a standard procedure will improve the security of the system especially during power system emergencies. To improve the system reliability it is advisable to keep records of information or instructions exchanged between man and man or man and machine. Documentation requirements have to be defined and met either manually or by using data logging machines, typewriters, hardcopy devices, etc.

8.3.1.2 The Information sub-system

Man-machine communication has to be coded so as to make the procedure as simple and effective as possible. The orders and instructions displayed to the operators must be clearly formulated allowing for easy and immediate interpretation. For specific information the operator has to select the required function and initiate the corresponding program. Security can be improved in the case of systematic or self-check information by a display of the instructions. This may be executed by the machine without operator intervention. It is also a matter of security that the personnel in charge of operation or maintenance of each component of the information system are made aware of the state of the other units. For example, it is advisable that the operators are informed when maintenance is under way on the communication equipment of a substation, or when part of the computers is under repair.

It follows that it is of outmost importance to pay attention to the definition of man-machine communication and the role of man in such a system.

8.3.1.3 The Executive sub-system

Two different situations have to be considered :

- *Manual control with the operator giving instructions by telephone to the stations.*

  The instructions and orders will be predefined as for the decision system. The instructions will be repeated by the station operator and actions will be acknowledged. Records of switching operations kept on special forms will improve security.

- *Remote control procedures*

  The device to be operated is first of all selected. A signal to the control centre will confirm that the correct device has been selected. Then the control instruction is set up, checked by the operator and transmitted. A signal shows the operator that the order has been executed. In case of a hierarchical structure, it is possible that the system enables one device to be controlled from two or more different centres. Then great care must be taken to ensure that two centres cannot control the same device at the same time. Either special instructions must be given to the operators or hardware or software interlocks must be provided which release the control facility under a given procedure.

8.3.2 Data Base

The principal operations concerning the data base may be defined as follows :

- Constitution: collection or definition of data, storing in memory and organisation.
- Analysis of the validity and accuracy of the data
- Modification and creation of data base.

For the data chronicles, the factors concerning the reliability of the data files are mainly :

- the validity and accuracy of data perhaps accumulated over many years using different data collection methods;
- the difficulties in obtaining coherent data;
- the large amount of data.

Data reduction methods (mean values, regressions, coefficients, etc.) can simplify the data files. In some cases it may be advisable to smoothen a series of values to eliminate those which are either singular or doubtful.

Organisation of these data files using a standard format and structure definition of the input routines of the various application programs enables the same files to be used by different programs.

The most secure arrangement is that which has one fixed data base common to all programs. However, this may result in a complex structure. The security conditions required for this may decrease the total security of the data system. Data files dedicated to particular programs may be created at any time necessary to maintain the flexibility of the system and the security of the application to meet the requirement of some special programs.

<u>Using the data base</u>

After all the security criteria for the creation of the data base have been defined and met, the system must

- meet the security conditions for the data base;
- give access to the proper data when they are called for by a program;
- have an access time compatible with the ECS operation requirement;
- avoid any disturbance to the data files.

8.3.3 Quality of Documentation

The documentation related to an ECS can be divided into two subsets:

- Description of the System;
- Operating regulations and procedures.

The various documents describing the system are of utmost importance for training the maintenance and operating personnel, for the modification and adaptation of the system, and to facilitate expansion.

The documents must be drawn up by the various specialists who design and implement the system. They must be comprehensive, clearly presented and meet the requirements of the application for which they are drawn up.

The quality of documentation must be based on the following main criteria:

- Presentation of the documents
- Source of information;
- Standard procedure for writing;
- Information retrieval;
- Communication of the documents to all the authorities concerned;
- Determination and respect of classification levels.

8.4 Equipment

8.4.1 Role and principal components of equipment

The relationships between the various sub-systems and with the power system can be represented in the following diagram (fig. 8.1).

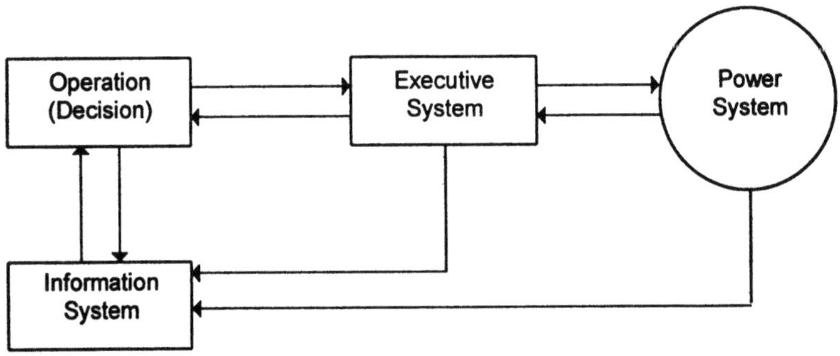

Figure 8.1: *Inter-relationships of sub-systems*

The major components of the equipments of an ECS embrace all the three sub-systems: operation or decision ; information; and executive.

In terms of equipment they can be divided into: metering and sensing devices, transducers, data communication devices, data acquisition and data processing equipment comprising computers, special purpose controllers, input-output units and peripherals, and display devices with man-machine communication devices.

The term "equipment" is extended to include the different type of software used in the computers: basic software (monitors, supervisors, compilers, data managers, etc.) and application software (operational planning, real time supervisory control, extended real time for security and economy).

The premises and buildings containing the equipment are also a part of the system and their design is also governed by the criteria of security and reliability.

For each component, security and reliability are defined by the quality of each device and by the structure of the sub-systems.

8.4.2 The Information System

The information system is split up into three parts: the telemetry system, the process computer system and the display system:

- Telemetry system comprising :

  DC: Data Capture (sensors, transducers).

  DT: Data Transmission : remote terminals in the stations, communication links (telephone lines, power line carrier, micro wave) telephone system.

  DR: Data Reception : information testing, validity, state estimation.

The accuracy of information reaching the Control Centres is influenced by the characteristics of each component of the telemetry systems, such as:

- the sensing devices: current and voltage transformers and transducers;
- the transmission into a coded form;
- the transmission (communication devices)

Transmission can be effected in either an analog or a digital form. In the latter case, the analog currents or voltages output from the sensors are coded after either sampling or integrating.

The final reliability and accuracy is given by the total availability of the chain and by the maximum error of the measurement reaching the control centre.

- Process computer system comprising :

  DP :Data Processing Computers: simple logical real time processing, extended real time with system solving calculations, off line computers for predictive functions; computer peripherals.

Experience of working with process control computers under real operating conditions points to the following ideal configuration:

> An on-line dual process control computer system, incorporating an automatic switch-over from one computer to the other in the event of failure.

The dual system works on the principle "two listen while one speaks". This method guarantees a fast on-line take-over by the stand-by machine. The switch-over has to be performed by hardware to avoid faulty logic being transmitted via the software.

Either single or two level structures are also possible.

In all cases the security conditions will be achieved by providing back-up arrangement at the same level or by accepting a degraded service.

– Display system comprising :

DD : Data Displays : CRT's , Mimic diagrams, recorders, typewriters, etc., man-machine communication facilities.

The process control computer serves purely as support for the control centre. It should prepare and condense the information so that a good overall view of network operation is available.

Particularly suitable for information presentation are the multicolour displays with graphic facilities. Suitable representation of the network structure (complete or part network diagrams with ZOOM and MOVE features, station layouts, etc.,) could eventually lead to mosaic wall diagrams becoming redundant. A large variety of items may be displayed on the CRTs e.g., diagrams plots, generator switching arrangements, tabulated readings, measurements, etc.

In addition, an emergency fall-back system should be provided capable of giving a minimum level of data and display facilities independent of the on-line processing system. Either separate hardware logic or the front end computers or even microprocessors may be considered for this duty.

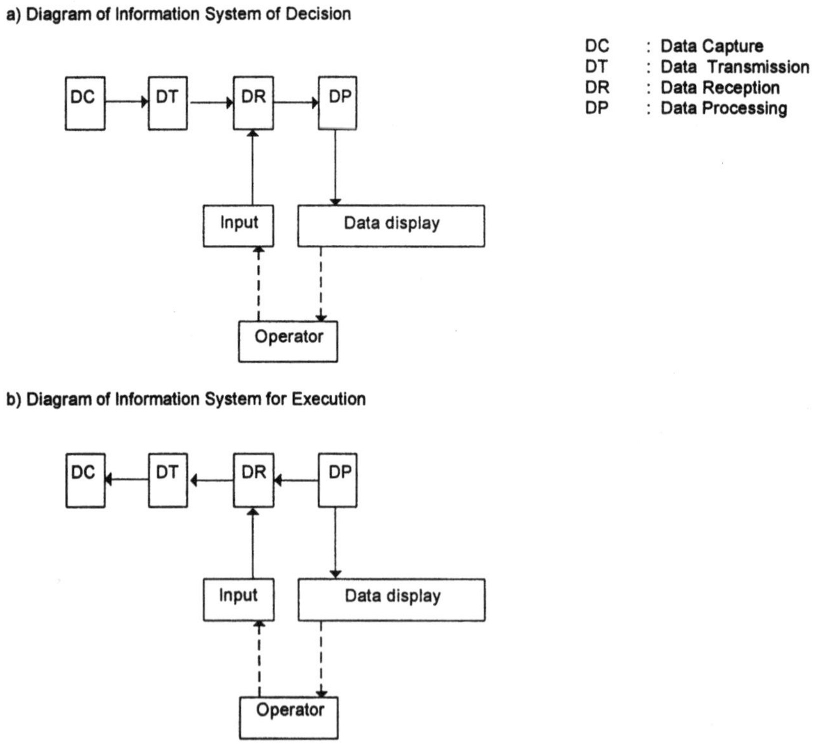

Figure 8.2: *Relationship of hardware components of the Information System*

Display facilities at the ECS should be engineered so that control equipment failures affect only a limited number of CRTs ; alternatively, suitable operational redundancy should be built into the system.

A representation of the relationship between the different hardware components of the information system is illustrated in Fig. 8.2

8.4.3 The Decision System (Fig. 8.1)

The process followed to ensure that the needs for security and operational economy of the power system are met consists of three phases of the decision system:

- long-term planning to define the equipment required, followed by a medium term analysis to test the operation conditions of the planned system;
- short-term forecasts in which generation, load flows, and power interchanges are defined;
- monitoring and control of the power system.

Assuming that the security and reliability conditions of the different types of equipment (mainly computers) have been met, the quality of these functions will depend on:

- quality and precision of the programs simulating these different phases of operation;
- consistency and accuracy of the data;
- functional and security criteria of power system operation.

8.4.4 The Executive System (Fig. 8.1)

The "executive system" incorporates all the rules and procedures and equipment and staff devoted to the direct control of the power system. It is the end result of the ECS and the part of the system requiring the highest level of security and reliability.

The executive system can be divided into three types of complementary sub-sets of functions and equipment :

8.4.4.1 Permanent Automatic Control

This includes such systems as Load Frequency Control, Economic Load Dispatch and Generation Control, Voltage Regulation, etc.

The security of this control depends on the security and reliability criteria of the information sub-system and requires close co-ordination with the local devices and the characteristics of the power units or network devices controlled.

8.4.4.2 Power System Protective Arrangements

Though not usually considered to be an integral part of the ECS, power system protection is a fundamental complementary subset of the executive system. It has a very important role in ensuring power system security.

Such protective equipment operates in conditions which are very different in terms of initiation, execution and response times from the action of the executive system's remote control. Action is initiated by detection of the contingency and executed within a fraction of a second.

For example, under-frequency relays, overload or fault detecting trip relays, network separation devices, automatic reclosures, etc., perform many actions which relieve the ECS operators and control devices of the need to carry out some emergency and other procedures to meet security requirements.

The security of these protective arrangements depends upon the type and position of the equipment, its setting, maintenance, general characteristics , response time and inbuilt redundancy.

8.4.4.3 Operating and Switching remote control equipment

Generally speaking, an automatic executive system requires first of all the transmission to the control centre of additional information to replace that given by the substation operator. The choice and quantity of information will somewhat condition the security of the executive system. Assuming the availability of this information, control can be executed with the following equipment levels:

- a simple communication "point to point" system similar to that of the manual system and requiring only average quality and security of each component;
- a system of computers which can be either specially devoted to remote control or the main data processing system of a control centre. The equipment configuration will depend on whether the ECS is centralised or decentralised;
- in a highly decentralised organisation, remote control will be executed from several master substations, each one having a small number of stations to control. In such a case the reliability required for each system is somewhat reduced, and in many cases such centres are equipped with only one computer;
- in a less decentralised organisation, control may be executed from the regional control centre, and in a centralised organisation from the Central Control Centre. In such cases the security of the computers and of the communication system is of utmost importance.

The man-machine communication equipment can have various features depending on the sophistication of the control, the degree of equipment centralisation and the complexity and size of the network to be controlled.

For smaller systems with or without computers, wall-boards with mimic diagrams and push buttons are very simple and reliable pieces of equipment.

For larger systems, with CRTs graphical substation diagrams can be displayed and individual plant items selected by using a marker. Selected devices may be indicated by the use of flashing or colour displays. An "executive" key allows the transmission of the order when the operator has checked the validity of the operation.

The marker may be moved on the CRT by using either keys, a light pen, a tracker ball or a joystick. Security of switching operations depends upon the ability of the operators to use these devices.

Whatever may be the system, the fundamental security requirement is that even if an operator makes a mistake or disturbances are generated by the equipment, its design should incorporate all the necessary checks and redundancy to prevent a switch maloperation.

8.4.5 Conceptual criteria for ECS

8.4.5.1 Reliability and security

The reliability and security of each equipment is of utmost importance for ensuring the proper functioning of the ECS. For evaluating the reliability of the complete system, it is necessary to define the influence of each element and determine the necessary reliability criteria of each sub-system. The security of the components may be improved by back-up which can be obtained in at least three ways:

- by having dual systems;
- by back-up between similar equipment in the same control centre;
- by mutual back-up between different control centres.

Any equipment required for the satisfactory and secure operation of the ECS will have a dependability which can be defined quantitatively.

A suitable parameter is the Mean Time Between Failures (MTBF) related to the failure rate λ by :

$$\mathrm{MTBF} = \frac{1}{\lambda}$$

Maintainability of equipment is given by its Mean Time To Repair (MTTR) which depends not only on the characteristics of the equipment but also on the maintenance organisation.

$$\mathrm{MTTR} = \frac{\text{total repair time}}{\text{total number of failures}}$$

Hence the availability as a measure of the "design availability for equipment comparison" is :

$$A = \frac{MTBF}{MTBF + MTTR}$$

8.4.5.2 Response time

The response time for the various functions to be realised is another important characteristic which should be taken into account in evaluating the performance of the equipment.

The response time will be considered as good or at least acceptable in terms of the operating criteria and is determined

- by an appropriate definition of the characteristics of the equipment (size of the computers, operating cycles, nature of man-machine communication devices), and
- by the estimation of the load of the equipment (volume of data, load on computers, access time and execution time of various functions).

The functional security of the system is evaluated not only in terms of the desired availability but also with reference to the response time expected for each function.

Assuming the reliability or failure rate and maintainability of each component, the reliability of the various sub-systems can be estimated by probability techniques.

8.4.5.3 Expandability and Adaptability of the ECS

By considering the likely development of the power system, operating analysis should be made in order to estimate the future functions to be developed to match the successive operation requirements.

A comprehensive design will help to ensure the adaptability and expandability of the ECS in terms of security, reliability and economy.

Basic criteria here are:

- the expandability should be considered for the various components of the ECS. Addition of remote terminals in the stations is a simple matter but data transmission and the processing of the extra amount of information will have other consequences for expandability
  - of the communication system (either more channels and / or higher transmission speeds);
  - of the computer system (input-output facilities, core storage and auxiliary memories and perhaps the addition of extra computers to the architecture);
- the expansion of the system would need the minimum rearrangement of the existing ECS and should only require the addition of complementary equipment. For this purpose system modularity may be favoured, as modification of one modular unit has only a minimum effect on the other

units;

- the expansion of the system should not reduce its reliability and security from that fixed during the initial conceptual stage;
- adaptation and expansion of the system should cause few or preferably no interruptions to the operation of the ECS, and the modifications should be carried out with minimum reduction to the security of this system in the meantime;
- no new additional or replacement units in the ECS should be commissioned without having been properly tested. If possible, the former functions and equipment configuration should be kept available in case of difficulties with new units;
- if possible, the modification period should be chosen outside the critical periods (such as peak periods) of the power system operation;
- the redundancies in the ECS designs (e.g. dual computers) should be used to develop the additional system units;
- if the redundancies are not sufficient, degraded configuration should be used with maximum security and work completed in the shortest possible time.

8.4.6 Main phases of implementation of an ECS

8.4.6.1 User's specifications

The procedure used to analyse the design of an ECS and the way the technical specifications are drawn up greatly influence the security and reliability of the system.

Generally the different phases leading to the placement of order for an ECS either on one main contractor for a turn-key project or on several manufacturers, are:

- <u>a feasibility phase</u> to determine several feasible solutions which are then analysed in terms of feasibility and cost;
- <u>a project phase</u> to define the organisation of the ECS, its functions and the type of equipment required together with the required reliability and degree of expandability;
- <u>technical specification phase</u> for preparation of a tendering document describing all the functions to enable the contractor or manufacturers to offer a system which fulfils the requirements of power system operation and which meets the conditions of security and reliability.

If the structure, precision and basic characteristics of the system and its equipment are not specified thoroughly, the manufacturers would offer their own solutions or equipment. But then the system may not suit the user's interest, and may lead to some confusion.

The specifications must also include all the services required (maintenance, training of personnel, supporting works, etc.) as well as the conditions of supply.

8.4.6.2 Development of software

This part of the ECS equipment is probably the most complex, giving rise to considerable problems. In particular, it is very difficult to estimate the time required for the development of software.

Ideally, the software package should be created by the user organisation itself. In this case, personnel with sufficient experience in software training must be available. This has the great advantage that from the very beginning the system is shaped to one's own interests and is essentially more secure, since the experts are continually available, even after the system's completion and commissioning.

In order that the software is completed and available for use on time, a well-defined organisation should be set up.

The actual programming must be preceded by a through analysis of all the functions. The eventual user must participate in these studies.

Particular attention should be paid to the security and reliability aspects of programming and documentation, since even after completion of the programs, the data and the programs have to be altered and corrected and not always by the same people who wrote them.

Extensive edit and test facilities (DEBUG, TRACE, etc.) increase system reliability and are essential for program development. Furthermore, it should be possible to implement new possibilities of the operating system without difficulty.

From the security point of view, the user should have the following possibilities:

- ease of entry of instruction by the user (e.g. program calls) and security against system failure resulting from incorrect input. For this purpose there should be a so-called HELP function (key or special command). This gives the user instructions in the event of an incorrect or faulty input;
- output should be quickly and easily observable;
- displays should show the information clearly and without overcrowding.

8.4.6.3 Auxiliary equipments

Whatever be the reliability of the main components of the ECS, for its overall reliability and security, the ECS depends on other auxiliary equipment necessary for its functioning.

The principal terms of auxiliary equipment are:

1) Power supplies for telecontrol, telecommunications and computers;
2) Air conditioning.

- <u>Power supplies for telecontrol, telecommunications and computers</u>

  The security of power supplies is governed by three main criteria:

  - permanency of supplies;
  - quality of protective devices;
  - adequacy of automatic switches.

To provide an AC supply of adequate quality at the Energy Control Centre (ECC), DC fed inverters should be used. The inverter system should be duplicated having auto changeover and bypass switching facility.

Incoming supplies to the ECC from the local electricity supply system should be arranged to be as secure as possible and should be backed up by an alternatively routed supply and an on-site standby generator (Diesel or similar). Where an alternatively routed AC supply is not practicable, a second standby generator should be provided.

Standby generator capacity to maintain the ECC fully operational should be provided. Fuel storage capacity to cover continuous operation for at least 10 days should be provided.

Automatic start-up and on-load switching facilities should be provided and be activated by high or low voltage, high or low frequency, or loss of supply. Manual start-up, shutdown and load switching should be provided from the control room.

- Air Conditioning

The electronic equipment and its component parts are manufactured to operate under certain conditions of temperature and humidity. To ensure reliability, the environment in which the equipment is located must therefore be regulated by air conditioning.

Most important are the rooms housing the computer and the telecontrol equipment and the control room itself.

If the building has a general air conditioning installation, the equipment and control rooms of the ECC can use the same system. However, a separate back-up equipment is advisable in case of failure or maintenance of the main system and to avoid the necessity of keeping the whole installation running during the night and on Sundays.

In all cases the conditioning equipment should include dust filters (removing particles of size $>$ 1 micron).

8.4.6.4 Premises and buildings

All equipment and structural design should be planned to minimise the effect of equipment failure on the ECS.

All precautionary measures should be taken for meeting the following situations:

- natural catastrophes (earthquake, etc.);
- acts of violence (sabotage, etc.);
- fire, internal technical failures.

8.5 Conclusions

An ECS is a very complex system which embraces large areas of responsibility. The analysis of the reliability and security of the overall system is also very complex. It is therefore difficult if not impossible to formulate a comprehensive strategy for secure and reliable ECS design.

The main components of the ECS discussed in this Chapter are obviously not independent of one another; it is necessary to examine their mutual influence and the impact of each on the quality of the whole system.

The basis of the analysis and most of the security criteria must be defined during the design and development stage of the ECS. Considerable emphasis is therefore laid on the analysis of this first stage. However, there are always differences between the actual operational environment and that anticipated at the planning stage which calls for reassessment of security after the system has been commissioned.

Experience of operation as well as the power system evolution is a factor which makes it necessary to visualise future development of the ECS and review of its security and reliability.

The overall security and reliability required of an ECS is a function of :

- the size and characteristics of the power system;
- the existing operational problems to be handled by the ECS;
- the structure and the functions of the ECS.

The main factors influencing the reliability and security offered by the system will be :

- the structure and organisation adopted for the ECS - in particular, centralisation or decentralisation of functions and resources significantly affect security and the available standby procedures;
- the role of man, his training, ability and discipline - this can be evaluated in terms of the level of automation of the ECS and the characteristics of the man-machine communication system;
- the structure, reliability and availability of the main equipment together with the organisation and efficiency of maintenance;
- last but not the least, the security of auxiliaries and premises.

Usually the choice of a given structure for an ECS and for the associated power system is decided by the management on the basis of economic, political, and geographic factors. The designer must take into account these basic factors to develop a system with maximum security.

Application of the various criteria applicable to various types of structure of organisation in the assessment of the security and reliability of a particular system, calls for adaptation to suit the characteristics of the particular system.

The effective adoption of security and reliability factors for the ECS will also depend upon the extra cost that the management is ready to meet for obtaining them. Economic consideration sometimes leads to the need to determine a series of degraded operating modes having an increased but acceptable level of risk.

References

1. UNIPEDE, *Criteria related to the improvement of Security and Reliability of Energy Control Systems*, First Draft - October 1977.
2. UNIPEDE, *Considerations relatives à l'amélioration de la sécurité et de la fiabilité des systèmes de conduite des réseaux électriques*, 40.2, 11-15 juin 1979.
3. E. Mariani and S.S. Murthy, *Control of Modern Integrated Power Systems.* Springer-Verlag, London 1997.

CHAPTER 9

INTER-SYSTEM EXCHANGES, TARIFFS AND BILLING

9.1 General principles of power exchanges

The primary object of interconnected operation of systems is the potential for utilising the available energy in the most economical manner and improving the level of system reliability. The underlying philosophy is that each utility has the primary responsibility of ensuring the highest reliability of power supply to its consumers at the lowest possible cost.

As a means of improving the economy of their operations, the utilities in an interconnected system declare moment to moment their cost of generation at various power stations. The highest cost power in each system is considered to be its incremental cost. Utilities that have surplus capacity at low cost generating stations offer it for sale to utilities that have a higher incremental cost, on the basis of sharing the savings. Generally speaking, the rate for such exchanges is the seller's cost plus half the saving on quoted rates between the two parties. This basic principle is applied to all economy transactions which are put into various categories depending upon their duration and the level of commitment on the part of the buyer and the seller regarding the purchase or sale of such power.

9.1.1 Programming

Each constituent Dispatch Centre of the Interconnected system draws up a daily schedule in MW on quarter-hourly or half-hourly or hourly basis, one day before the day of operation, by a certain fixed hour, say, 18:00 hrs, for each type of exchange and with every other constituent of the system. At this hour, all the characteristics (power, energy and price) of the exchanges would have been defined and the approvals of the concerned dispatch centres obtained. The daily schedules are then considered as formal programmes.

9.1.2 Frontier point

For each tie-line, the frontier point is defined by an agreement between the two constituents. If tie-lines are established between System A and System B, and between System B and System C, and if System A has to supply power to System C, it is necessary first to establish a programme of power exchange between A and B and a programme between B and C. The intervening system B generally levies a charge for wheeling power from A to C.

9.1.3 Inadvertent exchange

The inadvertent exchange is the difference between the algebraic sum of the metered total exchanges and the algebraic sum of the scheduled or agreed total net exchanges. The inadvertent exchanges are either paid for or compensated in kind, as may be defined in the agreement.

In this manner, the problems concerning commercial agreements are separated out from those of regulation of planned or unplanned exchanges. It is thus possible to develop many kinds of exchanges amongst the constituents, which could be superimposed simultaneously on one another while billing.

9.1.4 Billing

The billing for the power exchanged is done on the basis of 1) the daily schedule of each programme finally agreed upon between the load dispatchers, and 2) the total inadvertent exchange. The tariff for each kind of exchange is also defined in advance.

9.1.5 Bonus or penalty factors for regulation

The tariff structure can also provide for bonus or penalty factors for purposes of correct regulation of the frequency.

9.1.6 Modification of the programmes during operation

During the operational day, system conditions may change calling for modification of the programme. The programme as modified and agreed to up to the start of actual operation day is the final agreed programme.

In the event of a major disturbance taking place in the system or on the tie-line, the programme of exchanges may have to be modified after the incident by common agreement among the partners.

After the day of operation is over, the formal documents for each programme of exchanges, including the modifications carried out during the day as also the formal exchange of information concerning these final programmes are sent as final records to the billing and accounting section.

9.1.7 Control of exchanges

Each system will regulate the actual net total exchange with all the partners with respect to the programmed net total exchange.

The total inadvertent exchange of each system (i.e., the difference between the actual total net interchange and the finally agreed programme) is considered for billing. The inadvertent exchanges are computed for each partner. In UCPTE , for example, this is done by Electricité de Laufenburg in Switzerland.

When the day of operation is over, the concerned load dispatcher collects the data regarding actual exchanges from the tie point substation and establishes the diagram of inadvertent exchanges. This diagram is sent to the accounting and billing section for record. A typical diagram showing inadvertent exchanges of a system is given in Fig. 9.1

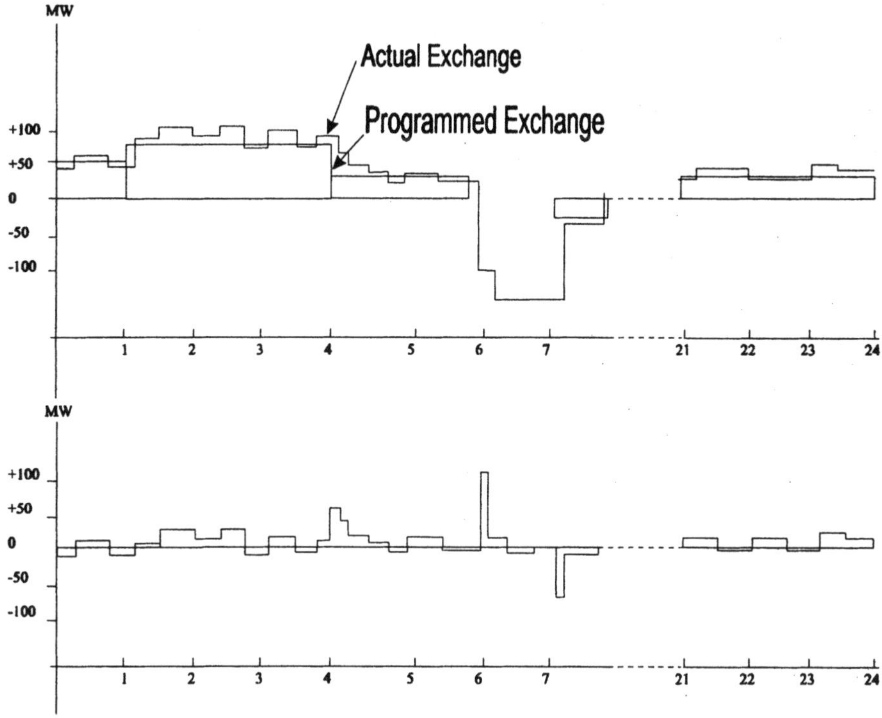

Figure 9.1: *Programmed exchanges, actual exchanges and inadvertent exchanges*

9.1.8 Tariff for inadvertent exchanges

There are two types of accounting for the inadvertent exchanges: Payment or Restitution.

(a) Payment : The rates are decided in advance and are generally different for peak and non-peak hours of the day and night hours. They may also be established with reference to the marginal cost of generation for all the systems put together in different months or seasons.

(b) Restitution: For this purpose the periods / days are classified as days, peak periods, non-peak periods, nights, Sundays and holidays. The inadvertent exchanges of each partner are algebraically summated according to the particular period. The algebraic sum of these inadvertent energies of all the partners for each period is always zero.

The inadvertent energies of various periods are regularly checked and when these quantities are sufficiently large, the partners decide to establish programmes of restitution to compensate for these differences.

If, for any reason, a partner finds it difficult to restitute the inadvertent energy drawn during peak periods, a price is established by common agreement between the partners for payment towards such inadvertent energy.

9.2 Power exchanges in the UCPTE countries

9.2.1 Pattern of exchanges

Power exchanges in the UCPTE countries broadly fall into two categories - long-term and short-term. Contracts are usually entered into only when the supply of firm power is involved, and the duration of these contracts is generally not less than five years. The two-part tariff is applicable for long-term agreements, whereas for short-term exchanges, only the proportional or the energy component charge is applicable, which is also dependent on the season and the hour of supply. A price variation clause is generally included in the tariffs with a view to providing for any variation in the economic structure in the respective countries.

While most agreements involve monetary payments for the energy exchanges, there are certain contracts according to which the country which imports energy at a particular time has to return the energy at some other time; for example, the country importing power during peak hours may be required to return three times the energy at off-peak hours. Similarly, a country importing hydro power in summer may be required to return only one-third of the imported energy from its thermal stations during winter season.

All the utilities maintain an hour-to-hour record of the costs of generation, transmission, etc., as also the incremental costs of energy production, which enables them to readily find out at which particular hour it would be advantageous for them to export or import power. In other words, transactions relating to power exchanges take place as in a stock exchange.

Apart from contractual exchanges, there are also "goodwill" exchanges. A further distinction is made as between neighbouring utilities and among non-neighbouring utilities.

An electricity exchange agreement concluded between non-neighbouring utilities A and C implies that the neighbouring utility B approves of this agreement and that B itself is not interested in this offer. The wheeling should not of course jeopardise the necessary security level or give rise to additional operational restrictions.

Due to the fast decisions implied in "goodwill" energy supplies, the agreements governing such supplies are primarily made between neighbouring interconnected utilities. If non-neighbouring partners are involved, the agreements are concluded on ad-hoc basis provided the intermediate partners have agreed in advance to the payment of wheeling charges by the end partners.

9.2.1.1 Energy exchanges and Accounting

The costing of power varies with the time of the day and is maximum when the demand is greatest. Fig.9.2 shows the tariff periods (low, high and peak tariff) as agreed between the countries under the UCPTE. Typical one-hour exchanges of energy within these tariff periods, governed by the contracts, among four UCPTE countries are shown in Fig. 9.3. The agreed imports and exports for the next day are fixed in a programme which then determines what generating facilities have to be employed. The system controller in turn supervises the flow of power over the tie lines and maintenance of the frequency.

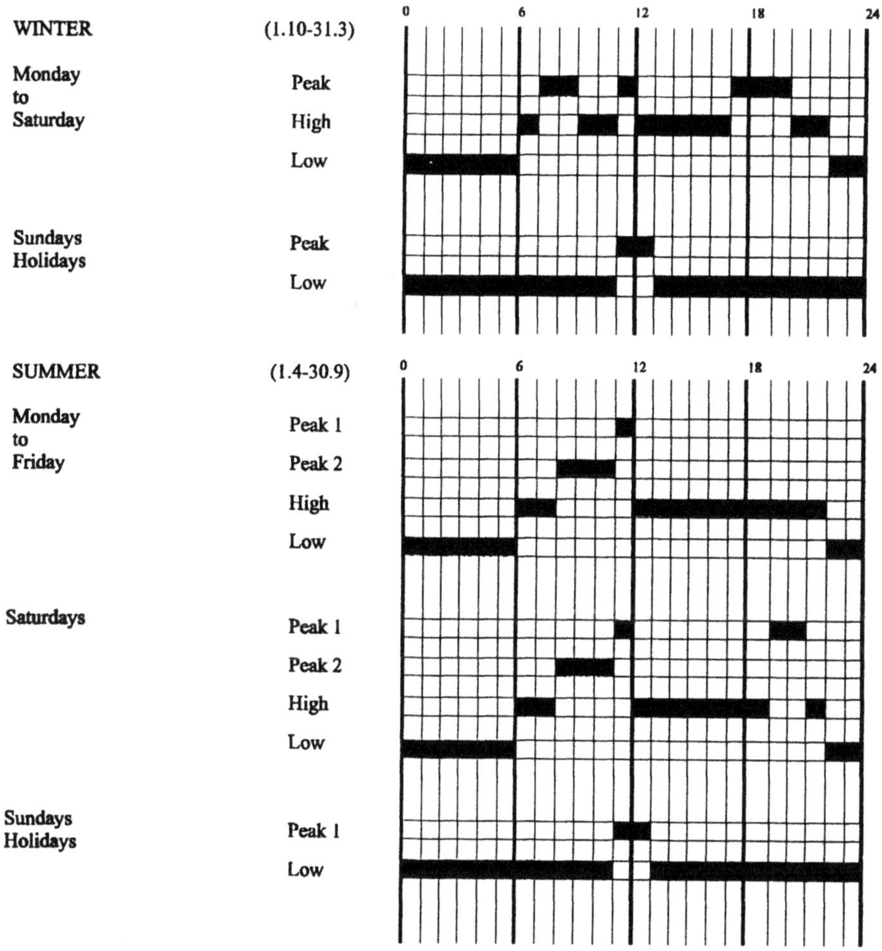

Figure 9.2: *Tariff periods in the European interconnected system*

A balance-sheet of all exchanges, over one hour or in some cases half an hour, is made out and it is the task of the accounting system to work out the difference between the agreed quantity of energy to be delivered and the amount actually exchanged within the tariff periods, for each of the partners (Fig. 9.3).

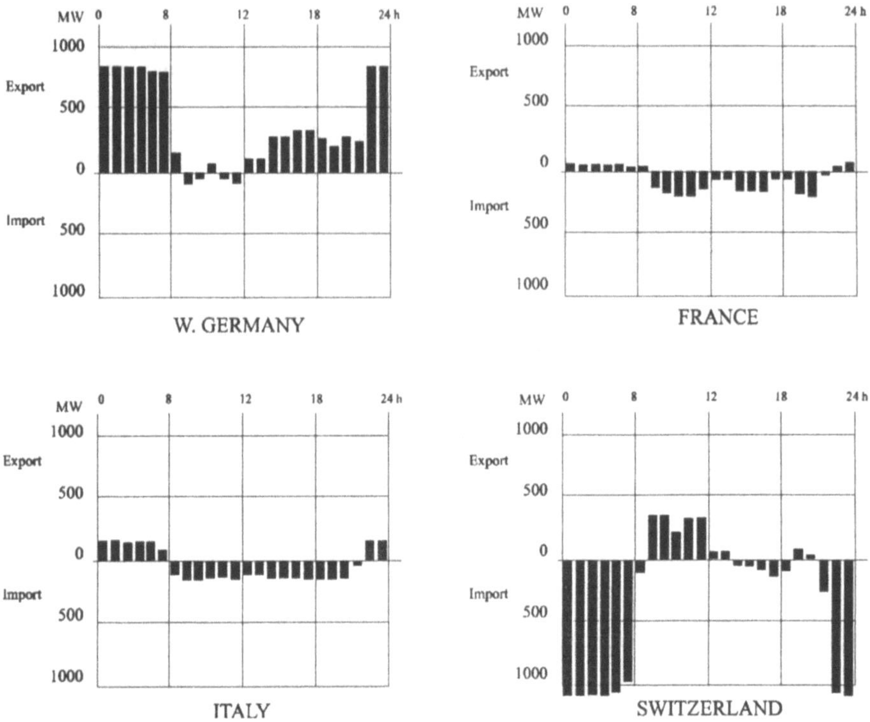

Figure 9.3: *Resultant transfer programme for the interconnected countries - France, Western Germany, Italy and Switzerland - as for 10 March 1965*

The accounting system has to satisfy that the tariff periods are adhered to, that allowance is made for the condition of system interconnection and that all the meters at the exchange points are read simultaneously (Fig. 9.4) Allowances have also to be made for the losses on the tie-lines during the period in question and the losses over the intervening or non-participant systems. The discrepancies - overdrawal or underdrawal - are then worked out for each system and each tariff period. To compensate for these discrepancies the accounting system establishes correction programmes additional to the control programmes worked out from the planned exchanges of energy.

The purchase and sale of electrical energy is based on the figures provided by meter readings. Due to inadvertent deviations (irregular consumption, faults, etc.) there are bound to be certain discrepancies from one undertaking to another, between the desired and the actual quantity of energy exchanged. These discrepancies are determined hourly and are corrected suitably.

Given the structure of the major energy resources in the UCPTE countries and the United Kingdom (with which France is interconnected through a submarine cable) as given in Table 1 and the scope for optimisation, the exchange pattern between France and the neighbouring countries is shown in Fig.9.5.

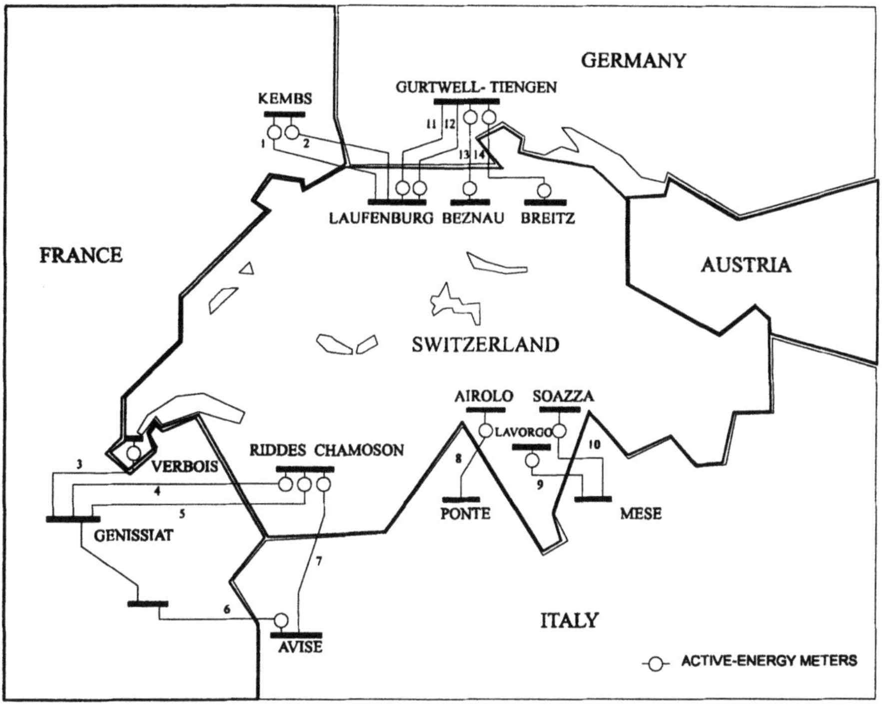

Figure 9.4: *Measuring the energy transferred*

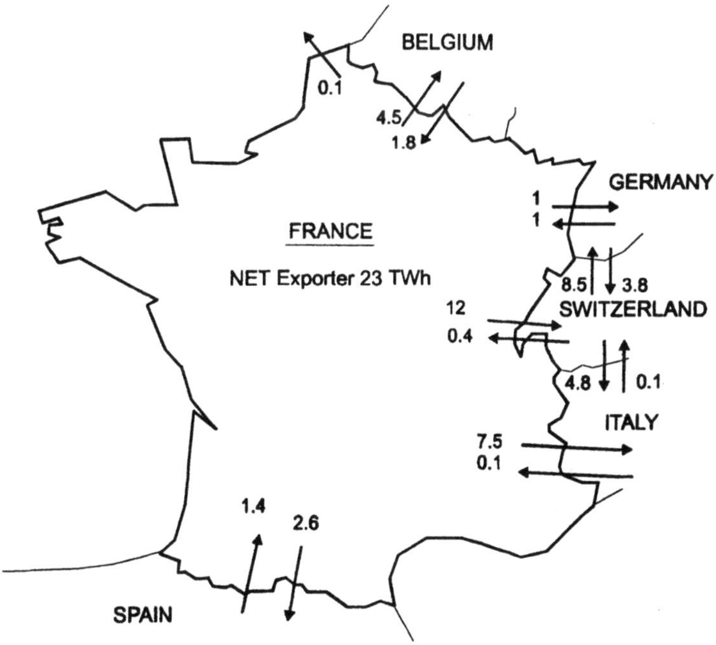

Figure 9.5: *Cross-frontier energy exchanges in 1985 (TWh)*

Table 1: *Pattern of major energy resources (%) estimated in 1990*

Country	Nuclear	Coal and Lignite	Hydrocarbon	Hydraulic
Belgium	57	29	12	2
Spain	28	37	6	29
France	76	4	3	17
Italy	8	21	51	20
Holland	5	35	60	0
Switzerland	39	2	2	57
Germany (West)	34	53	9	4
United kingdom	26	66	5	3

Fig. 9.6 shows, for example, how the nuclear energy available in France could be exported to the neighbouring systems.

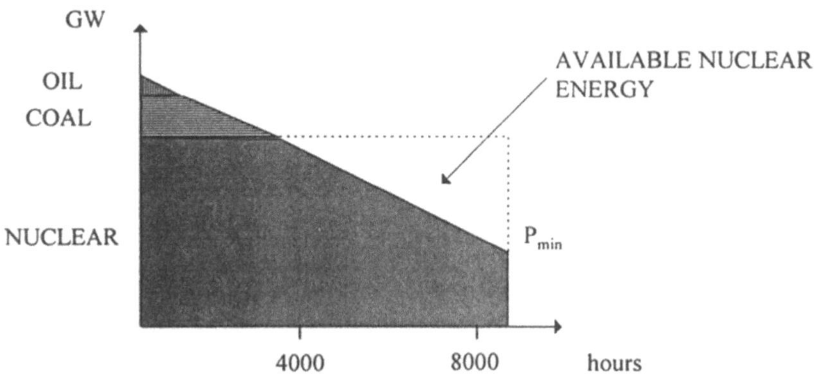

Figure 9.6: *Annual load-duration curve and optimum utilisation of thermal power*

Similarly Fig. 9.5 shows how Switzerland imports energy from France and exports to Italy and Germany, although not at the same hour.

The interconnections at 400 kV level between France and the neighbouring systems are shown in Fig. 9.7.

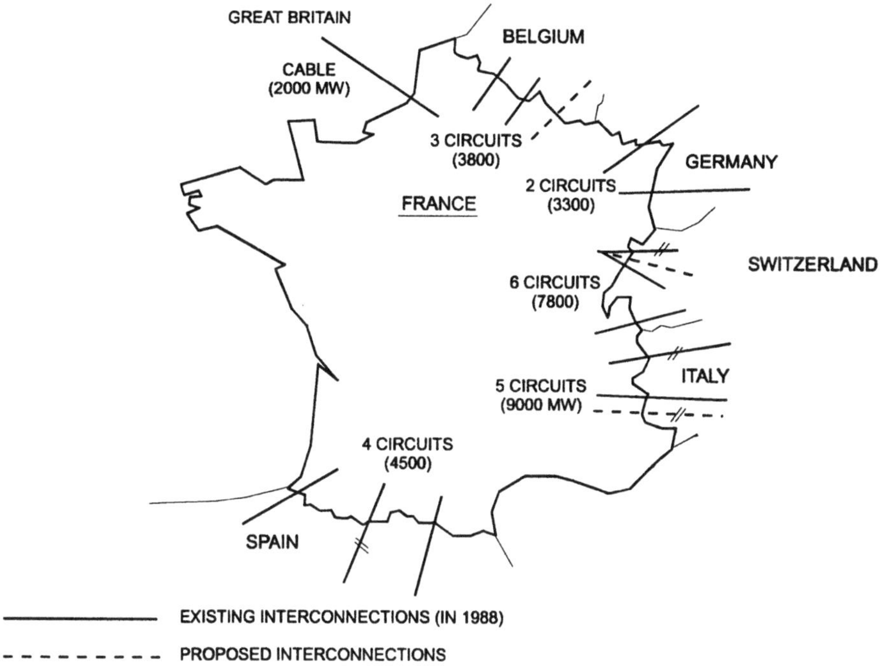

Figure 9.7: *Interconnections of France with the neighbouring countries at 400 kV (Thermal limits in MW)*

9.3 Power pooling and interchanges in the USA

9.3.1 Operating hierarchy

There are approximately 3500 utilities in the USA categorised by ownership and nature of operations under four distinct segments: Investor-owned utilities; State, municipal, and other non-federal publicly owned agencies; Co-operatives; and Federal agencies. The total installed capacity is around 700,000 MW. In spite of their large number, diversity of ownership and widely dispersed geographical area, there is considerable intersystem co-ordination to secure the maximum possible power efficiency and economy.

The nine Regional Reliability Councils that were created following the Northeast Power Failure of 1965 have become increasingly important in providing mechanisms for co-ordination of operating practices and planning criteria by all the bulk power supply utilities within the broad geographic areas in the USA. The Councils develop voluntary standards with a view to providing regionwide reliability of service.

The North American Electric Reliability Council (NERC) which was founded in 1968 provides the necessary co-ordination of policy issues among the Regional Reliability Councils and evolves recommendations at the national level on the reliability and adequacy of bulk power of the power utilities in North America.

Each utility is responsible for meeting the entire load in its own area of supply, and exercises system control individually for its own power system, with extensive co-ordination with the neighbouring utilities. The operating hierarchy consists of the following levels :

(i) Control Areas

One or more utilities may together form a single control area for purposes of exercising control over the generation, power system operation and tie-line exchanges. Each area is operated as per established norms for regulating generation, frequency, voltages and tie-line flows.

(ii) Power Pools

A number of contiguous control areas normally have operating agreements to provide mutual assistance under various types of operating conditions.

(iii) Regional Reliability Councils

These Councils are a co-ordination mechanism to provide broad operating guidelines and a forum for review of the actual operation of the interconnected systems in their respective areas.

(iv) North American Electric Reliability Council (NERC)

NERC is a forum for evolving national guidelines and discussion of operational issues relevant to the entire North American power system.

The key to the successful interconnected operation of a large number of systems with the total installed capacity in a single interconnection being as large as 350,000 MW, is the clear definition of control areas and strict observance of agreed operating rules and procedures. The following ten principles that define the major responsibilities and requirements of the utilities in a control area as laid down by NERC, are given below :

I. Each control area is required to provide sufficient capacity to carry its expected load at normal frequency with provision for adequate reserve and regulating margin.

II. Each control area is required to provide accurate and reliable automatic tie-line bias control as a means of continuously balancing its generation against its load, so that the net loading of its tie-lines agrees with the scheduled net interchange, plus or minus its frequency bias obligation.

III. Each control area operating in parallel with other control areas is required to have its frequency bias set equal to its area frequency response characteristic (Darrieus condition).

IV. All interconnections are required to be equipped with tie-line telemetering to the appropriate power control centres for inclusion in the area control schemes. Common measuring equipment is required to be used by both parties.

V. All interconnections are required to be equipped with kilowatt-hour metering, with readings obtained hourly at the power control centres as a means of continuously monitoring control area regulating performance and effecting prompt corrective action.

VI. Co-ordination of changes in scheduled power deliveries between control areas is essential for proper frequency control so as not to burden other control areas or systems.

VII. In the event that a system is deficient in either generating or transmitting capability, that system has the responsibility of bringing load and generation into balance within emergency limits.

VIII. Whenever practicable, generating units are required to be operated in such a way that governors are free to respond to changes in system frequency.

IX. Operating instructions and procedures are required to be established by each system to cover its operation under emergency conditions, including the loss of communications.

X. For high degree of service reliability under normal and emergency operations, it is essential that adequate and reliable communications be provided within a system, between systems and between control areas.

It has been possible to achieve this rigorous system discipline through comprehensive contracts and formal agreements only, as neither the Regional Reliability Councils nor NERC have any statutory or contractual authority.

9.3.2 Typical interchange transactions

The power pooling arrangements are purely voluntary and based on commercial agreements between the parties. The benefits of interconnected operation are, in all cases, fully quantified and equitably shared between the parties.

The co-ordinated economic dispatch of electric systems in the USA may be broadly classified into three groups :

- Central dispatch
- Sequential dispatch
- Broker dispatch

9.3.2.1 Central dispatch

About 40% of the total installed capacity in the USA comes under nine large centrally-dispatched groups. The utilities in the PJM (Pennsylvania- New Jersey-Maryland) power pool were among the first to form central dispatch pools. They share the savings from central dispatch and payments are arranged after-the-fact. The collective generation is dispatched based on incremental production costs. The central dispatch centre continuously receives the costs for each generating unit. The pools have free-flowing transmission. The additional costs of equipment, and operating and accounting personnel, are shared between pool members.

9.3.2.2 Sequential dispatch

This is the most common method for co-ordinating exchanges among utilities and approaches a free market for exchanges. Each utility has its own dispatch centre which exchanges information with neighbouring dispatch centres. Sequential dispatch involves bilateral and multilateral arrangements.

The following are some typical capacity and energy transactions:

- Firm capacity exchanges denote the degree to which a buyer can rely on the availability of capacity and its associated energy and are paid for daily, weekly and monthly tariffs based on negotiated prices. Likewise, the associated energy is a negotiated price of cost plus 0 to 15%.
- Economy energy exchanges are usually interruptible with very little notice and are generally priced halfway between the seller's incremental cost and the buyer's avoided cost of producing that energy.
- Economy capacity and energy exchanges are used as a last resort when the buyer is unable to supply capacity and energy for all of its customers. They are generally priced based on typical energy costs from very expensive peaking units plus 10 to 15%.

9.3.2.3 Broker dispatch

In brokering, each participating utility, after arranging for most of its exchanges using sequential dispatch, posts its quotes for buying and selling additional economy energy each hour. The buys and sells are than matched by a simple computer program to maximise the savings from the resulting transactions. The broker matches the highest-priced buyer with the lowest-priced seller until there are no more buyers or sellers or until the difference between the buyer's and seller's costs are less than a specific amount (usually $2-4/ MWh). The computer printout notifies the utilities of the amount and price of each transaction it has matched. The utilities then make the bilateral transactions if they choose to.

Each of these methods has a profit incentive in-built in the exchanges. The sale will not occur below the seller's cost and the purchase is not more expensive than what could have been produced by the buyer.

Both sequential dispatch and brokering make for a very competitive environment and allow sharing of cost information. The central dispatch group, however, does not share such information with outsiders. Exchanges outside the central dispatch group are made sequentially.

Reliability is paramount with all utilities in North America no matter which dispatch method is used. However, centrally-dispatched groups provide an overview of the entire area and are thus able to optimise the reliability and economy of the group. They are able to respond to emergencies more quickly and use the facilities of the group more effectively than with the other two methods. A major effort is now being made to enhance information exchange through communication-computer links with a view to improving reliability co-ordination for all utilities.

The various types of interchange transactions in one typical power pool, viz., the Central Area Power Co-ordinating Group (CAPCO) comprising five utilities in the North Central Region of the U.S., are given below for illustration:

Emergency Power - Energy and capacity supplied at short notice to enable one party to meet its load, operating reserve and firm commitments.

Economy (A) Power - Energy supplied by one party to the other on an hour-to-hour basis to reduce the costs of operation. The rate is based on the seller's cost plus one-half of savings on quoted rates between the two parties.

Economy (B) Power - Energy and capacity supplied by one party to the other on a day-to-day basis. Since capacity is implicit in this transaction, this service carries a mutually agreed time notice before cancellation. Savings are shared as for Economy A.

Short-term Power - Capacity and energy provided for periods of one day or more, usually one week or more. These transactions are used to supplement the generating capability of the receiving party or as an Economy purchase.

Limited-term Power - Similar to Short-term power except that the period of reservation is for one or more months.

Non-Displacement Power - Energy which is not covered in any other agreement between the parties. Receiver has ample capacity.

Fuel Conservation Power - Capacity and Energy scheduled on a weekly basis to meet a shortage of fuel, making it necessary for the deficient party to conserve energy sources over an extended period of time.

Unit Power - Specified amount of capacity out of the portion of a particular CAPCO unit owned by the supplying party for a specified period of time.

Backup Power - An obligatory transaction between parties within CAPCO to provide capacity and energy to a party who is deficient because of the outage of CAPCO units or because a party cannot meet its minimum reserve requirement.

The list indicates the extent of co-operation between these utilities for enhancing system reliability and reducing operating costs through better utilisation of resources available to any utility in the short term.

The long-term measures for co-operation include the establishment of jointly-owned capacity which allows the setting up of large size units while permitting a phased investment programme for the individual utilities. One utility, in whose area the power station is located, is designated as the operating agency and operates this capacity on behalf of all the partners and is held accountable to them for its operation. The shares of each utility in this capacity are operated as per the requirements of each of them and may even be sold to others, within or outside the power pool. Other long-term arrangements include purchase of capacity or firm blocks of power from utilities that have generation in excess of their own medium and long-term requirements. Such capacity is treated as part of the resources of the purchasing utility.

9.3.3 Scheduling, Billing and Accounting

9.3.3.1 Scheduling

Scheduling of power exchanges between systems has assumed considerable importance as it forms the basis of numerous interchange transactions which take place. Power scheduling has therefore emerged as a separate discipline as distinct from dispatching. In fact, the Bonneville Power Administration (BPA) in the USA has power schedulers on duty round the clock.

Power scheduling is not limited to operational planning only; it is also a vital aid in actual operation and enables prompt billing in respect of power exchanges. Normal billing is based on schedules, and metering only identifies the deviations to be settled either in kind or through minor monetary adjustments.

9.3.3.2 Pricing

Pricing of intersystem transactions is not related to the base costs considered by the utilities in formulation of tariff for their consumers. The rates for exchange are intended to cover the incremental costs only. The concept of out-of-pocket Cost adopted by some utilities is representative of this general philosophy of pricing in respect of interchange transactions and is briefly explained below.

The out-of-pocket costs of supplying power in each hour are the costs incurred in the supply of the highest cost power available on the supplying party's system during that hour.

The components of these costs are the following:

<u>Operating capacity costs</u>

- Start-up and shut-down costs
- No load cost
- Maintenance cost
- Charge (or credit) for increased (or decreased) cost of energy generated by the Party associated with the transaction
- Incremental labour costs
- Applicable incremental taxes
- Miscellaneous incremental operating costs

<u>Energy Costs</u>

- Incremental fuel cost
- Incremental transmission losses
- Incremental labour cost
- Incremental maintenance cost
- Applicable incremental taxes
- Miscellaneous incremental operating costs

The Federal Energy Regulatory Commission (FERC) must approve pricing for all inter-utility exchanges. The FERC will generally approve prices which are not higher than the seller's investment-related and operating-related costs. Utilities often charge less, however, in today's highly competitive spot-market for electricity. Considerable flexibility and ingenuity is used as sellers and buyers compete to arrange exchanges.

9.3.3.3 Billing for interchange transactions

Billing in respect of interchange transactions is based on the net scheduled exchanges in various categories. Each system compares its scheduled exchanges, in each category and with different utilities, with a net telemetered flow on all its tie-lines, on an hour-to-hour basis. The scheduled exchanges and metered tie-line flows are then compiled on daily and finally on a monthly basis for preparation of the net interchange account.

The difference between net schedule and net metered exchange is treated as an inadvertent flow to be adjusted between utilities with corresponding surpluses and deficits by suitably modifying the schedules at times when load conditions are similar to those when the deviations actually took place. The difference between the billing meters and the data compiled from telemetered signals results in a cumulative error which also has to be taken into account by suitably adjusting the scheduled exchanges over a period of time.

Commercial settlements in respect of interchange transactions are made once a month. It takes generally 8-10 days for the billing to be done. The time permitted for settlement of bills is normally of the order of 15 days.

The commercial aspects of such transactions such as rates, energy accounting and billing are important because their accomplishment with reasonable speed, accuracy and reliability is necessary for the free flow of power and interconnected operation of the power systems.

9.3.3.4 Accounting Practice

The utilities concerned normally ensure that the data pertaining to actual and scheduled hourly interchange, utilised by them for the hourly power accounting functions is identical. In order to ensure accurate data, common watt-hour meters, accumulators and hourly freeze signals for the remote stations are utilised.

A uniform system of accounting is adopted by all utilities which greatly simplifies the task of the regulatory agencies in reviewing and analysing performance with regard to the criteria laid down by them. As most of the interchange transactions are based on the members sharing the savings, it is important that the costs of each utility in relation to such transactions are well defined. The adoption of uniform accounting practice enhances the acceptability of cost data furnished by each utility in respect of the various interchange transactions.

9.3.3.5 Wheeling of power

The interconnected operation of power system is dependent upon the free flow of power across the boundaries of utility systems, power pools, States and even national boundaries. Utilities in an interconnected system are obliged to provide wheeling service subject to the availability of capacity on the bulk transmission and related facilities during the period for which wheeling service is required.

In view of the large magnitude of power exchanges to the mutual benefit of all interconnected systems, wheeling of power is viewed as a service provided by a utility to other interconnected systems on a no-profit no-loss basis. However, the incremental costs legitimately allocable to such transactions are recovered.

The actual flows over the tie-lines are of a smaller magnitude as power flows, in actual practice, take place by displacement. The availability of telemetering and modern data processing facilities has made it possible to use on-line load-flows to determine the incremental impact of each transaction on the transmission system losses.

9.4 Banking transactions and Barter deal

Two interesting features of inter-state energy transfers sometimes followed in the Southern region of India are the Banking transactions and the Barter deal.

9.4.1 Banking transactions

The System which is in a position to bank, called the Banking system, banks spill hydro or surplus thermal energy with a system deficient in energy, called the Banker system. In case there is spilling in the banker's reservoirs (when banking operation is in process), the energy first spilled is considered to be the banked energy and no liability accrues to the banker system due to the spillage of banked energy.

9.4.1.1 Accounting of the banked energy

Banker being surplus

The banker system returns the energy banked in its reservoirs after accounting for the following:

I. 5% of the energy banked towards evaporation losses in the reservoirs.

II. Transmission losses in the system (transmission losses shall be reckoned in both directions) when the energy is sent into the banker system's reservoirs for banking and also when the energy is returned by the banker system.

III. Service charges equal to 5% of the cost of net energy banked shall be levied on the banking system. In case the banking system is also a surplus system (which does not require the banked energy to be returned), the service charges shall be borne by the deficit systems which share the banked energy in proportion to their shares.

In case there is spilling of all the banked energy, no liability accrues to the banking system. If the spilling is partial, the evaporation losses and the service charges are applicable to that part of the energy which has not spilled.

Banker being deficit

The banker system shares the energy banked in its reservoirs with the banking system in the ratio of 20 : 80 after accounting for the following :

I. 5% of the energy banked towards evaporation losses.

II. Transmission losses in the system reckoned both ways.

However, in case there is spilling of all the banked energy, no liability accrues to the banker system. If the spilling is partial, the evaporation losses and the sharing of banked energy are applicable to that part of the energy which has not spilled.

9.4.1.2 Tariff for banked energy

The tariff for the banked energy from systems which are entirely hydro is the weighted average cost of generation at the stations where spill would have occurred. For energy banked by systems which have both hydro and thermal generation, the tariff for banked energy is the pooled cost of generation (including hydro and thermal energy) in the banking system.

9.4.2 Barter deals

These refer to transactions involving release of commodities like cement or water for irrigation purposes in lieu of power supplied by a surplus system to a deficit system.

9.5 New organisational structures

9.5.1 Structural organisation of the functions of the electricity supply industry

An electricity supply utility usually has a combination of generation, transmission and distribution facilities (the so-called "vertical" organisation). In recent years, the process has started of "unbundling" these facilities, that is, generation, transmission and distribution are owned and operated by different companies or utilities ("horizontal" organisation); in this case a co-ordination of facilities is often performed by an ad hoc organisation.

As a particular case of generation utility, there are the so-called "independent power producers" (IPP), and "non-utility generators" (NUG). For example, industrial undertakings which have their own electric power generation for their internal use, and sometimes also produce process steam ("combined heat and power", CHP), sell the surplus power to, say, a transmission utility; these are commonly called NUGs. Another case is that of a commercial undertaking which builds an electric power station just to produce and sell electrical energy, in competition with other undertakings; such undertakings are commonly called IPPs.

9.5.2 Ancillary services

A vertically structured utility has the main task of producing, transmitting and distributing electrical energy to the customers, but it also takes upon itself the accompanying task of performing a set of "ancillary services" like regulation of frequency and voltage. Such ancillary services naturally involve technical and economic considerations. A particular example is that of modulating the active power generation in accordance with the variation of system demand during the day.

It appears therefore quite natural or even necessary to involve NUGs and IPPs in such operations -- apart from the generation utilities in the case of horizontal structure. This responsibility should be made clear in the regulations defining the relationship between generation bodies and, say, a transmission utility receiving power from them.

NUGs in particular have however the natural tendency to avoid the responsibility of providing such ancillary services, and to confine themselves to the task of delivering active and reactive power at, for example, a constant value all through the 24 hours of the day. This tendency is due to the fact that companies playing the role of NUGs are often more interested in following the electrical generation schedule required by their internal processes than the schedule appropriate to the service of transmitting and distributing electrical energy to external customers.

9.5.3 Third party access

As mentioned above, there is the case of an IPP which offers to market its electrical energy at a competitive price; there is a transmission utility which will transport this energy; and there is a customer, who is generally located geographically far away from the IPP, willing to buy the energy utilising the service of transportation performed by the transmission utility. This is the so-called problem of "third party access" (TPA), that is, the problem of giving access to an IPP to a transmission facility and of giving to a customer the possibility of buying energy from a producer of his choice. The problem is similar to that of wheeling. Of course, this process requires to be regulated. The choice is generally limited to consumers whose annual energy consumption is above a particular level (for example, the domestic consumers are excluded from this possibility).

9.5.4 Security of interconnection

As far as the security of interconnected operation is concerned, some regulations include, for example, a clause that, if the frequency of the interconnected system decreases to a predetermined value, or in case of overload of tie lines, the tie lines themselves are opened, the systems separate from each other; this is done in order to prevent the propagation of heavy disturbances from one system to the other.

The frequency limit at which the separation occurs should be co-ordinated with the load shedding plan of the interconnected systems.

Other regulations include the requirement of monitoring periodically the short-circuit levels in the stations near the frontiers, and making periodically some static security calculations on the interconnected system to check that the power exchanges are within acceptable limits and the maintenance plans or schedules comply with the exchange plans.

9.5.5 Marketing of electrical energy

In general, there are no regulations in regard to overall optimisation of the interconnected operation (economic aspects); an optimisation is indirectly performed on the basis of the free market competition, since it is supposed that each interconnected utility knows, from moment to moment, its own incremental generation cost reported at the electrical border line; if c_1 and c_2 are the incremental costs of the networks 1 and 2 respectively, and if $c_1 > c_2$, it is natural that utility 1 buys energy from utility 2, for example, at a price $(c_1+c_2)/2$, which is lower than c_1 (advantage for utility 1), and higher than c_2 (advantage also for utility 2).

It is assumed, as is the common practice in North America, and begins to be so in Western Europe, that the sale prices of electrical energy related to incremental costs, and corresponding amounts of available power and energy (e.g. hour by hour) of each partner are known to all the other partners; that is, there is "transparency" in commercial transactions. The broadcast to each partner of the interconnected system giving the sale prices and quantities of available power and energy could be done by means of a dedicated telecommunication network, similar to the practice adopted in stock exchange markets.

This is the scenario for the so-called short-term or "spot" contracts of exchange, which cover the next day or the next week. These contracts may be discontinued, at short notice from the selling to the buying utility, if the need arises. There are also medium- and long-term contracts (firm contracts), covering months or years, which are stipulated on different grounds (for example, on the basis of long-term availabilities of energy from nuclear generation, or from pluriannual hydro reservoirs).

For short-term contracts, the assumption is that the cost of energy is different in different periods of the day. According to the UCPTE practice, the day is divided into "peak hours" (from 7 a.m. to 9 a.m., from 11 a.m. to 12 a.m. and from 5 p.m. to 8 p.m.), "full load hours" (from 6 a.m. to 10 p.m., except peak hours) and "slack hours" (from 0 to 6 a.m. and from 10 to 12 p.m.). This division applies to a working day and is different for other days and varies with the season.

9.5.6 Current status of application of new structures and functions

At present the processes of unbundling and the TPA are resorted to by certain governments partly for political reasons but mainly with the object of introducing competition in the power sector.

Till a few years ago, and even now in many cases, the power sector was considered a "natural monopoly" and hence exempt from the concept of competition; the utilities, vertically structured, had (and have) a concession from Government in a given territory, and were (and are) subject to more or less strict regulations on tariffs. In other words, it was considered natural that, in a given territory, the production, transmission and distribution of electrical energy should be performed by a single agency, in order to avoid the duplication of some structures, especially transmission and distribution networks, etc.

As of now, there is no general agreement on the new structures and functions mentioned above; several experiments are under way. In Europe, there is the case of the United Kingdom, with its horizontal structure deriving from the unbundling of the former vertically structured electrical industry, and with TPA function active; and there is also the case of Spain, again with a horizontal structure. In both cases the transmission utility covers the entire national territory.

References

1. Luder, H., *The Interconnection of Swiss Power Networks as part of the West European System*, Load Dispatching, Brown Boveri Publication (undated).
2. Albouy et al., *Planification des interconnexions internationales L'approche d'EDF*, Revue Générale de l' Electricité, no.7 - Juillet 1988.

APPENDIX 1
PRESENT PRACTICES IN LOAD FORECASTING

PRESENT PRACTICES IN LOAD FORECASTING

This Appendix describes load forecast techniques based on the results of a survey of 44 utilities carried out by CIGRE in 1991.

1. Introduction

Load forecasts play a predominant role in many decisions related to electricity supply systems. Without reliable demand forecasts, the other electricity system optimisation tools, however excellent they may be, cannot yield accurate results.

2. Time Spans

Load forecasting is generally divided into the following time-spans:

- long-term (5-30 years), econometric forecast which plays a fundamental role in economic planning of new generating capacity and transmission networks.
- Medium-term (1 month - 5 years), which is used mainly for the scheduling of fuel supplies, maintenance programmes, financial planning and tariff formulation.
- Short-term (1 day - several weeks) , which provides the basis for planning start-up and shutdown schedules of generating units, reserve planning and study of transmission constraints.
- very short-term (some minutes - several hours), to deal with economic load dispatching and security assessment.

3. General Characteristics

The survey covered 44 utilities from 18 countries as under :

- Western Europe	52 %
- Asia	21 %
- Australia, New Zealand	14 %
- America	13 %

- Utility Characteristics

The size of the utilities, in terms of annual energy, varies between 1.6 and 333 TWh, the average being 44 TWh.

The annual peak utilisation time varies from 4464 h to 6824 h, the peak utilisation time being obtained by dividing the annual energy by the annual peak demand.

The ratio between peak and valley demand for the same day averages to 1.59 in winter and 1.66 in summer. These average figures have large variations; for instance, in winter the ratio varies from 1.08 to 2.4 from one utility to another. The ratio is higher during winter in the case of 35% of the utilities while it is higher in summer in the case of 55% of the utilities; the remaining 10% of the utilities have the same ratio for both winter and summer.

While the load breakdown varies from utility to utility, a typical breakdown is as follows:

- industry 50%
- residential 30%
- commercial 20%.

4. Objectives of load forecasting

The objectives of load forecasting are listed below, along with the percentage of positive responses from the 44 utilities:

	MLT	ST	VST
Unit commitment	34	80	30
Loading of Units	27	75	64
Optimisation	66	82	68
Reserve Planning	72	73	34
Power and energy contracts	67	52	14
Study of transmission constraints	65	68	25
Capacity planning	88	18	0
Planning of fuel ordering	82	14	0
Financial planning	85	0	0
Others	7	5	2

The main objectives of medium / long term (MLT) forecasting are planning of capacity, financial and fuel ordering.

The short term ST load forecasting is helpful in unit commitment, loading of units, optimisation and spinning reserve.

The primary justification for the very short term (VST) load forecasts is that they are used to adapt the short term generation scheduling for optimisation.

These forecasts are also useful for scheduling power exchanges between utilities and study of transmission constraints.

5. Parameters influencing load forecasts

5.1 Main Parameters

Load forecasts are more or less influenced by factors related to the kind of consumers and their sensitivity to external factors. The importance of these parameters is not, however, the same when the forecast horizon changes. So, while weather variation effects may be forecast in the short and very short term, the economic situation is a source of major uncertainty for the medium / long term.

The following table shows the relative importance in terms of percentage of the responses, of the main influencing parameters according to the forecast horizon:

	MLT	ST	VST
Industrial load variations	63	50	23
Weather variations	16	93	61
TV broadcast events	0	27	39
Others	11	18	11

As the major parameter is that of weather variations, a closer study of this parameter is useful.

5.2 Meteorological Factors

Temperature is the main meteorological factor considered in load forecasting as 95% of the utilities take that into account. Another important factor is the extent of cloudiness: 50% of the utilities take account of this factor. Wind is also taken into account by 27% of the utilities.

As regards the temperature, its influence on the demand varies not only between winter and summer, but also between the peak and the valley of the same day. About a third of the utilities which are sensitive to the winter/summer temperature have a higher load gradient in summer than in winter.

The gradients are summarised in the following table:

% MW /°C	Winter Peak	Winter Trough	Summer Peak	Summer Trough
Average	1.26	1.13	0.64	0.18
Std. Dev	0.85	0.93	1.80	1.59
Min	0	0	0	0
Max.%	3.70	2.8	4.0	4.4

As regards load variations according to temperature, most utilities use thresholds, ranging from 10 to 18°C for negative gradients (space heating) and from 22 to 30°C for positive gradients (air-conditioning).

In most cases (80%) weather forecasts are supplied from a state agency; some utilities (11%) use a private weather forecasting agency and 20% of the utilities have their own weather forecasting service which are for the most part used to confirm or further refine the forecasts coming from outside sources. Some utilities use more than one option.

5.3 Special Events

Few utilities are able to supply significant load variation values due to television. As a rule, these remain less than 1% of the total power consumed. This may well be the result of the diversification of programmes and the multiplication of channels. Some utilities have indicated that the most difficult demand forecasting is associated with bank holidays, industrial strikes and public events.

6. Methods

6.1 Load Forecasting in Practice

To summarise, in practice a large number of forecasting procedures developed by electric utilities appear to be a hybrid of several types of statistical procedures, especially in the case of short and very short-term load forecasting.

Although all utilities are confronted with very similar situations, the response reveals a wide spectrum of approaches. The methods range from relatively simple procedures to complex algorithms.

6.1.1 Techniques

In Table 1 a summary of the different techniques used by different utilities is presented. A short description of these techniques is also given.

Table 1: *Summary of Different Techniques*

	Multiple regression	Spectral decomposition	Exponential smoothing	Kalman filter	Box - Jenkins	Others & Hybrid
MLT	14 (32%)	5 (11%)	3 (7%)	2 (5%)	9 (20%)	6 (14%)
ST	15 (34%)	1 (2%)	0 (0%)	1 (2%)	5 (11%)	10 (23%)
VST	7 (16%)	0 (0%)	2 (5%)	0 (0%)	5 (11%)	8 (18%)

Many utilities did not complete this section of the questionnaire apparently due to the techniques not being widely understood.

Multiple Regression

This is perhaps the most widely used technique and one of the simplest . It uses the classical linear regression formulation in the form :

$$L_i = B_i + \sum_{j=1}^{n} (K_{ij} \times W_j)$$

where :

L_i = Load demand in hour i,

B_i = Base load, or part of the load not dependent on weather variables.

W_j = Weather variable j

K_{ij} = Regression coefficient

n = Number of weather variables

This model is normally used for off-line forecasting and the base load and regression coefficients are updated using other techniques like exponential smoothing. The main advantage of this method is the simple manner in which the weather influence on load can be introduced. One disadvantage is the fact that this formulation ignores the autocorrelation of the time series that can produce jumps and discontinuities in the forecasting activities.

Box - Jenkins (Time Series Analysis)

This technique analyses the structure of the load time series. It is at first modelled as an ARMA (autoregressive moving average) process composed of autoregressive terms AR(p) that represent the random component of the variable value of the present hour as a weighted average of p past values.

$$Z_t = \Phi_1 Z_{t-1} + \ldots + \Phi_p Z_{t-p} + a_t$$

and moving average terms MA(q) that represent the contribution to the random component as functions of weighted averages of q deviations a,

$$Z_t = a_t - \varnothing_1 a_{t-1} - \ldots - \varnothing_q a_{t-q}$$

Since the ARMA models apply to stationary processes, the non-stationary processes should experience a differencing operation. If the order of this difference is d, then the process results would be modelled by an "autoregressive integrated moving average" model ARIMA (p, d, q).

Once the load series is modelled, if external variables are to be included, a Transfer Function is used. This transfer function is a linear model that sometimes does not reflect accurately the weather influence on the load.

One disadvantage of the Box-Jenkins model is that the parameters are not to be updated on-line.

Exponential Smoothing

The basic hypothesis for the validity of this method is that the evolution of the demand is regular, that is, abrupt variations are not present.

The forecast is based on "deseasonalised" values of load, seasonality factor and trend, as follows:

$$L_d(m) = A\frac{L(m)}{F(m-s)} + (1-A)\left[L_d(m-1) + R(m-1)\right]$$

$$F(m) = B\frac{L(m)}{L_d(m)} + (1-B)\cdot F(m-s)$$

$$R(m) = C\cdot\left[L_d(m) - L_d(m-1)\right] + (1-C)\cdot R(m-1)$$

where:

m is the present time

$L_d(m)$ = deseasonalised value at time m

$L(m)$ = observed value of load at time m

$F(m)$ = seasonality factor at time m

$R(m)$ = trend at time m

s = seasonality cycle

A, B, C are exponential smoothing factors, with values between 0 and 1 and chosen "a priori" on the basis of some criteria. Low values of these factors give more weight to old values of L, F and R and hence provide a more stable behaviour of the process; high values give more weight to recent values and hence make the model more responsive to the dynamic or changing behaviour of the load evolution process.

The forecasted value of load at time m+t is then:

$$L(m+t) = \left[L_d(m) + t\cdot R(m)\right]\cdot F(m-s+t)$$

This method is used also for on-line updating of off-line prepared model parameters. The main drawback of this method is the determination of the smoothing factors.

Kalman Filtering

Kalman Filtering is a statistical estimation technique of recursive nature in which the estimation of the state in a moment is a result of a weighting operation between the extrapolated past values and the value of the current observation.

This approach is suitable for on-line application, but it is necessary to elaborate the model off-line.

Spectral Decomposition

In order to analyse the time series two general approaches have been developed: one on a time domain (finite parameter, Box-Jenkins), and the other on a frequency domain; these two can be considered as complementary techniques. It is in the last domain in which the Spectral Decomposition technique can be used.

The underlying concept in this analysis is that each variable or function over time can be meaningfully represented by pure sine waves summed up over different frequencies with different amplitude and phase at each frequency. Performing this analysis for pairs of variables, the degree of linear association between them at different frequencies can be determined.

6.1.2 Off- and On-line Systems

In general the forecasting systems may be distinguished as off-line and on-line systems.

Whereas off-line methods often employ regression models in which the load is related to exogenous variables, such as meteorological factors, on-line forecasting procedures adapt to the recent telemetered demand values, by the use of time-series techniques as there are spectral analysis, exponential smoothing and Box-Jenkins methods.

In the on-line facilities an important determining factor on overall performance is the quality of the input data that derives from the telemetered information. Such data is often subject to both steady state random errors and occasional intermittent errors. This requires sophisticated methods for data validation.

In some cases an interactive prediction is applied, in which an automatic system allows human intervention to enter off-line information into an on-line predictor. This can be an elegant way to deal with meteorological factors.

Besides the methods mentioned in the above table, utilities have mentioned econometric models for medium-term load forecasting, pattern selecting and adaptive prediction in very short-term load forecasting.

A popular device in on-line short / very short-term forecasting seems to be the standard load curve approach.

It is often produced on a daily basis, reflecting long-term average behaviour. The residual deviations between the actual load and the standard load are computed and adaptively forecast. A strong point mentioned in its favour is that it very readily involves the co-operation of the operation management.

6.1.3 Current Performance

About 60% of the utilities are satisfied with the performance from their current methods. Those who are not satisfied mentioned the following reasons (in order of importance) :

- unacceptable errors in terms of power
- would like to include weather effects
- would like to automatise the method
- the model is hard to build
- unacceptable errors in terms of energy
- the model is not easy to maintain
- handling of special days
- the model degrades easily

6.1.4 Weather variables

Weather variables are often a subject of discussion in load forecasting models. Their effects on the demand of electricity have been analysed in many (case) studies.

The importance of temperature, windspeed, cloud covering and humidity often depends on the composition of the load : for example, if air-conditioning is a substantial part of the load. Therefore, results of studies often cannot be used without a case study for the particular situations of the utilities.

Implementation of weather variables requires quite accurate weather forecasts or in case of on-line predictors a reliable weather metering system.

7. Data and hardware

7.1 Characteristics of the data

The essential data for the accurate load demand forecasts is the availability of recorded load data. These data should comply with two conditions:

- Data should be highly reliable
- The amount of data must be sufficient for the purpose of the study.

The first point appears evident, but sometimes this fact is not properly considered. The quality of the data used is a factor of paramount importance. No one load-weather model can work properly if the data used for its building or tuning does not represent the behaviour of the variables to be modelled. This is specially true in the case of load data. These data should be carefully filtered from anomalous data which in many cases are very difficult to detect and to be corrected. This effect happens quite often due to reasons like strikes or holidays that affect just a part of the load served, and consequently sometimes the impact of such contingencies on the load is very difficult to evaluate, and in such cases it is wise to discard the load data.

The quality of the weather data is also as important as that of the load data, but in this case there is normally no difficulty in obtaining good weather data from their own utility records or from the National Meteorological Organisation which keeps records of many weather variables in an international standardised format for many stations.

It is important to always keep in mind that the model to be built will be as good as the data used, and that the use of more sophisticated mathematical models in order to capture the individual effects of the weather variables for the different load patterns, increases the necessity of better data.

On this subject , and in the light of the replies received , it appears that 40% of the utilities keep records of contingency corrected hourly load time series, i.e., corrected from the impact produced by special events, strikes, special holidays, etc. The importance of maintaining such data need hardly be emphasised.

Almost all the utilities keep the data in the form of hourly load data. There are a few that store the information in the form of half-hourly load data. The use of quarter-hourly load data and five minutes load data which is quite unusual, is also reported.

Half of the total number of replies received report that a series of weather-corrected load are also recorded. The use of this kind of series has a drawback in that the series have to be recomputed in case the load-weather model changes. This affects mainly the short and very short forecasting activities, whilst it is not so important for medium and long term. In most of the cases the weather correction refers to the temperature, and in a few cases this correction is made using other weather variables as cloud covering or wind.

7.2 Maintenance and Collection of Data

As regards the manner in which the load information is collected and stored in files, nearly 70% of the utilities have in operation some kind of (or partly) automatised procedure for this purpose. About 40% of the cases report on the maintenance of this data base more or less manually.

7.3 Period considered

While planning to create a model for load forecasting, an important question is to determine the amount of historical data to be used, or in other words, how far in the past should the data be taken in order to capture all the characteristics of the relationship between the load and other variables like weather variables, etc.

The duration of the period to be used should be evidently greater in the medium-long term than in the case of short-term. In the case of the medium-long term the past period should be long enough to be able to model the seasonal patterns of the demand, but not so long that the most recent dynamics of the demand are not represented properly.

- 34% of the respondents use between four and seven years of data for this purpose.
- 16% use ten years of data.
- 16% use more than ten years of data.

As far as the short-term is concerned the duration of the load data series used by the different utilities lies in the range varying from a few years to one week.

7.4 Computational Means

The manner in which the forecasts are prepared and the computational environment in which this task is performed has also been an object of analysis.

In regard to the medium-long term, 43% of the utilities use computerised procedures installed in an off-line environment. Just 5% of the respondents use the on-line computers for this job.

As regards the short and very-short term, the number of utilities using procedures installed in the off-line computers is about the same as the number using the procedures working into the on-line environment.

In the short and very-short term it is very convenient to have the activities of model updating and forecast inserted into the real time environment. From the survey results, it is apparent that 70% of the utilities lack this possibility. Furthermore, many of them report that their future efforts will be addressed to implement such procedures.

8. Medium-long term

All the forty-four utilities state that they perform medium or long term forecasts. This notion, long term, is however a bit vague as regards the forecast horizon, 20% of the duration terms exceed 5 years (there are even some that reach 30 years) and refer rather to capital investment choices than those related to operation.

The most frequently duration encountered for the medium-long term forecast is 5 years (20%), however, some prefer 3 years and a few 1 year.

With such a horizon, sectorial forecast techniques prevail over global forecasts; this can be explained by the availability of efficient forecasting tools in this domain and by the interest of examining structural and time-related distortions of electrical power consuming activities. As far as the energies are concerned the forecast is made either month-by-month, per week or even per day. As for power, about one third of the responses indicate hour-by-hour forecasts. The others prefer to limit themselves to the peaks, the energy and valleys either for each day, the week, the month or even the year.

As regards the medium-long term forecasts, about 35 % the respondents state that they examine several scenarios concerning the annual growth of their demand. The standard number of scenarios examined is three and this most probably means that the scenarios considered are base, high and low.

A posteriori controls of the forecast quality for annual and medium term horizons are made by 4 out of 10 utilities. The forecasts are more accurate for energy than for power.

For the annual horizon, energy errors vary between 1 and 3%, depending on the utility, but power errors reach 5% or even higher. One of the responses indicated that use of an error formula of a+bt type, where t is the forecast horizon given in years.

9. Short-term

Almost all the utilities make short term forecasts; the most frequently encountered horizon is 7 days (55%), but longer, up to 10 to 15 days, and shorter, 1 to 2 days, are also indicated.

The basic short term forecast interval is most of the time (3 out of 4 utilities) 1 hour; others, however, use shorter time intervals : 1/2 or 1/4 hour.

It appears that when there are both short and very short-term forecasts the time intervals are the same as the methods and models are usually identical.

Consequently, the power forecasts always correspond to the elementary calculation intervals. It should be noted that the daily energy forecast is a concern for 1 out of 3 utilities and only 9% of them concern themselves with the energy forecast for the week.

This is in fact consistent with the methods used as 68% of the responses indicate forecasts made directly off the load curves without having first made an energy forecast.

There are load curve forecast variants, some utilities mentioning that peak and valley forecasts are a pre-requisite to determining the hour-by-hour loads.

It would appear that some 66% of the utilities are able to estimate the forecast errors according to the horizon; the corresponding figures most frequently given for a 24-hour horizon are from 1 to 4% in power and from 1 to 3% in energy.

10. Very short-term

Almost all the utilities feel that the load forecasts have to be updated during the day. The horizons fixed are quite varied, ranging from 5 minutes to 2 days. However, more than half the responses indicate either the end of the current day or a sliding 24-hour period.

While most of the time the forecast interval is the same as that used in the short term forecasts, for some 20% of the cases the intervals used are equal to or less than 15 minutes; figures given by utilities having high activation recurrences which are usually automatic. It is possible that these utilities use very short term forecasts within the framework of Automatic Generation Control (AGC).

However, 50% of the utilities update load forecasts at operators' request, requests due to major differences observed between the forecasted situation and that which actually occurred or as a result of updated weather forecasts.

Power forecast errors, with a horizon of several hours, are around 1%.

11. Related issues

11.1 Share of load forecast among load buses

For many studies it is not sufficient to know the total load of the whole system or even the total load of a region. In general when it is necessary to consider the transmission network, the load demand at every load bus should be determined.

From the analysis of the replies received on this aspect it is observed that :

- 40 % of the respondents considered no division among load buses.
- 50 % of the cases perform some kind of load forecast for load buses. 18% of them do it just for medium-long term; 5% just for short-term; and 23% of them for medium-long and short-term. The rest use coefficients based on statistics for distributing the load among the load buses in a post-processing activity.

11.2 MVAr Demand

We have been considering so far the forecast of the demand of active power, but what happens to the demand of reactive power?

Do the utilities consider the MVAr demand forecasting an important issue?

Perhaps these are questions difficult to answer in a general manner. Every power system has its particular characteristics and consequently its own operating constraints due to the flow of reactive power.

The replies received can be summarised as follows:

- 85% of the replies reported that no MVAr demand forecasting activity is performed.
- 15% of the replies show that some kind of action is taken in this regard.

11.3 Monitoring the Errors

Load forecasting quality is based on three points :

- The Model : this is the representation, usually formalised, of the load and parameters influencing it : temperature, wind, type of day, season, economic activity.
- The Method : this is what makes it possible to reduce the observed load to "normal" values and to extrapolate to the future firm forecasts made on the basis of the evolution of the parameters.
- The Data : this is the information, observed or forecasted, by which one knows past load and the values of the parameters.

These three points have to be studied in a consistent manner. Any study of forecast errors must have as its initial objective the determination of the relative weight of each of these points within the overall error.

So, by simulating a load forecast, using actual meteorological condition, one can simultaneously find out the gains made by the improvement in the forecasting of these factors and the residual error related to the model and the method.

Similarly, the data time period has to be optimised in terms of the forecast horizon so as to determine which are good explanatory parameters and what is the best relationship between load and parameters.

To sum up, the improvement of load forecasting requires the development of extended a posteriori analysis.

12. Users' opinion

Many utilities state that they are not satisfied with the load forecasts and even among those satisfied, many have some reservations. Most of the criticisms concern the short and very short term forecasts, probably because the inaccuracy of these forecasts is immediately seen.

The main criticism concerns weather forecasts, either they are deemed to be of insufficient quality (40% of the cases) or because the forecast models do not take account of them. Note that more than 60% of the companies perform post event analysis on weather forecasts. Several utilities indicate problems during holiday periods and believe that an experienced operator is just as efficient as an ARIMA type technique.

Some criticisms relate to the application of various techniques and would like the very short terms models to have the calculations available on-line.

The survey showed that all the 44 utilities prefer load forecasts with emphasis on the short and very short term forecasts.

The variety of methods used reflects the diversity of utility and demand structures. Any system used must reflect the need of various operational and planning activities.

13. Conclusions and recommendations

A perfect load forecast is an impracticable target and, in any case, probably too expensive. Each utility has to analyse the price it pays for forecast errors, either on the basis of its own experience or by making simulations with its decision aid tools.

Some of the topics that could be profitably considered are :

- load forecast error and sizing of the spinning reserve
- load forecast error and sizing of fuel stock
- load forecast error and cash management
- load forecast error and maintenance planning of generation and transmission equipment.

Two approaches can be suggested for these studies:

- Compare the management cost expectancies for a random future (with the present forecast errors) with those for a known future (no error). This will yield an upwards factor for the overall risks.

- Study the gains brought in by a qualified, reasonable improvement in the quality of the forecast (e.g. a 20% reduction of the short-term temperature forecasts' standard deviation). This will give the stakes sensitivity to the measure that are easiest to implement.

Of course, these expected gains have to be compared to the outlays required to improve the forecasts.

The occurrence of numerous non-typical situations often makes the algorithmic forecasting methods inefficient, requiring corrections by the operator.

References

1. IEEE Committee Report, *"Load Forecast Bibliography, Phase I"* IEEE Trans. Power App. Syst., Vol. PAS - 99, No. 1, pp. 53-58, 1980.
2. IEEE Committee Report, *"Load Forecast Bibliography, Phase 2"* IEEE Trans. Power App. Syst. Vol. PAS-100, No.7, pp, 3217-3220, 1981.
3. G Gross and F D Galiana, *"Short Term Load Forecasting ; A Review"* Proc IEEE, Vol. 75, No. 12, pp. 1558 - 1973, Dec. 1987.
4. D W Bunn and E D Farmer, *Comparative Models for Electrical Load Forecasting*, John Wiley & Sons Ltd, 1985.
5. M A Abu-El-Magd and N K Sinha, *"Short-term Load Demand Modelling and Forecasting; A Review"*, IEEE Transactions on System, Man, and Cybernetics, Vol. SMC - 12, No.3, pp. 370 - 382, May / June 1982.
6. E Moline, I L Ayuso, *" Integrated System for Power Demand Forecast"* , CIGRE 1988 Session, Paper 39 -16.
7. S Rahman, R Bhatnagar, *"An Expert System Based Algorithm for Short Term Load Forecast"*, IEEE Transactions on Power Systems, May 1988.
8. Kun-Long Ho et al, *" Short Term Load Forecasting of Taiwan Power Systems using a Knowledge-based Expert System"*, IEEE Transactions on Power Systems, November 1990.

APPENDIX 2

A-E

(REFERENCE CHAPTER 6 - LOAD-GENERATION BALANCE)

APPENDIX A

1. Partitioning of time in the scheduling problems

Especially for hydro-thermal power systems, scheduling is an activity in which the decisions taken today for tomorrow or for the next week can affect future operation requirements. Hence the optimum way to schedule is to consider as long a period as possible (for example, up to one or five years ahead). This may be true also for purely thermal power systems, for instance, in the case in which there are strict constraints in the consumption of contracted fuel within a given period (say, one year).

Nevertheless, the data required to plan in detail the operation for tomorrow or the next week is more than that for the guideline of operation of next month or the next year.

Keeping in mind this fact, and the problems related to complexity and the length of computations, it appears wise, in addition to other simplifications to be assumed in the representation of the power system itself, to consider at least two levels of time partitioning :

- the first one is of more direct interest in the medium-long term operation planning, in which the period considered (for example, one year) is divided into n "medium-term time intervals (MTI)"; for example, the duration of each MTI might be one week, and n=52 or 104 or more;
- the second is of more direct interest in the short term scheduling, in which each of the above defined MTI is divided into m "elementary time intervals (ETI)", each of duration, typically, 0.5 or 1 or 2 hours.

In the following it will be assumed for the sake of simplicity that recourse is made to this two-level partitioning of time.

2. Simplifications

It is assumed that, as far as medium-long term operation planning is concerned, simple power system models or equivalents may be considered (for example, only one equivalent, or aggregated, hydro reservoir will be considered; transmission constraints will be neglected).

Nevertheless, the basic ideas and methodologies remain valid when applied to more complex problems.

Some data, often considered as known (e.g. load and water inflows, available generators), are in practice random and not deterministic. To account for randomness, which is particularly important if the period of study is long, requires further sophistication of the models.

When the short term scheduling period is one week, randomness may still have some importance, especially as far as the last days of the week are concerned. Nevertheless it may be noted that those utilities having a short term scheduling period of one week or more repeat the short term scheduling procedure e.g. every day, so considering operational, in practice, only the next (or the next two) days schedule .

Hence in the short term scheduling, randomness is normally accounted for by making available an appropriate amount of spinning and standby reserve (computed perhaps with probabilistic criteria, Chapter 5).

3. Categorisation of generation, load and power exchanges

For any duration, and in particular, for any MTI, the generation must match the load plus exports (average values along the time interval considered, MWh or GWh).

The generation may be made up of hydro and thermal sources (including nuclear if any and peaking power such as gas turbines).

Hydro may be split into two terms: H, coming from reservoirs, and H', corresponding to the water which may not be stored during the time interval under study.

In a similar manner thermal generation may be thought of as being composed of T, which may be varied, and T' ("must run" at full load or at the maximum available load: for example, geothermal and mine-mouth plants; in some utilities nuclear generation may also belong to this category).

The load may be considered as being composed of a firm load L' (that is, the load regulated by rigid contracts) plus a secondary load L (load regulated on an interruptible tariff).

In some utilities there may not be any distinction between the two categories of load, in the sense that all load is considered as primary (that is, L=0); the interruptible customers, if any, are in this case considered as firm load which may be curtailed if given prior notice.

Exports may also consist of an amount E' corresponding to firm contracts ("must sell"), plus an amount E, corresponding to opportunity trading. The same may be said for imports I' (" must take") and I.

4. Constraints

In the i-th MTI the following energy (GWh) balance will hold good:

$$H + H' + T + T' + I + I' = L + L' + E + E'$$

Let now be:

$$F = L' + E' - I' - H' - T' \quad (1)$$

the amount of firm energy, that is, the energy subject to strict commitments, or not subject to optimisation.

Also let

$$M = L + E \quad (2)$$

the amount of secondary load plus opportunity export, and finally let,

$$U = T + I \quad (3)$$

the variable thermal generation plus opportunity import. Note that M, U, L, E, T and I are all non-negative values.

The first equation may now be rewritten as:

$$H_i + U_i = F_i + M_i \text{ for any i} \quad (4)$$

Equation (4) shows a balance with one fixed (forecasted) value F, and three values H, U and M which may be varied, to some extent, by the operational planner.

Now, let us define A_i (MWh) as the inflow to the equivalent hydro reservoir, and V_i as the storage (MWh) at the end of the i-th MTI; the following equation will hold good:

$$V_i = V_{i-1} + A_i - H_i \text{ for any i} \quad (5)$$

which represents the reservoir balance. An additional constraint is

$$0 \le V_i \le V_{max} \text{ for any i} \quad (6)$$

where V_{max} is the maximum storage.

In general there will also be some other constraints of integral type, such as:

$$a_{min} \le \sum_i H_i \le a_{max} \quad (7)$$

$$t_{min} \le \sum_i T_i \le t_{max} \quad (8)$$

Also H_i and T_i have upper bounds H_{imax} and T_{imax}.

In (7), in which a_{min} and a_{max} (GWh) are specified and known amounts of hydro reservoir generation, it may often be $a_{min} = a_{max} = a_{tot}$; in such a case a_{tot} could, for example, equal the total amount of inflow to the reservoir (that is, the sum of the n values A_i), so that (7) would impose on turbine this total inflow within the whole period considered; as a result, it would be $V_n = V_o$.

Similarly, (8) states that the total thermal generation is constrained within a given range t_{min}, t_{max}, due, for example, to constraints imposed by fuel purchase contracts.

5. Criteria of optimisation

The three variables H, U and M, subject to (4), (5), (6), (7) and (8) must be set at values which minimise the operation cost for the period under study.

It should be noted that the presence of eq. (5) (that is, the presence of hydro storage), as well as the presence of the in equations (7) and (8), make the problem an integral type one with respect to time; hence the setting in a particular MTI may not be determined independently of the settings in the other intervals.

In other words, the actions taken for interval i affect the economy of the following intervals, and hence the economy of the whole period (which is composed of n MTIs).

Various cases will now be considered, as far as generation mix and market are concerned.

5.1 First case : no hydro storage (or "purely thermal")

Note that the meaning of "no hydro storage" is strictly correlated to the duration of the MTI; if this is equal, for example, to 2 hours, "no storage" means the absence of ponds and reservoirs, or the presence, at the most, of-run-of-the-river hydro plants, or no hydro plants at all.

If the MTI is equal to one week, on the contrary, "no storage" means the absence of monthly or seasonal reservoirs (but there could be daily and weekly ponds).

To clarify this point, it should be noted that the following assumptions have been made: a daily pond shall have the same storage at the beginning and at the end of the day (it can make transfer of generation from hour to hour, within the day); a weekly pond shall have the same storage at the beginning and at the end of the week (it can make shifts of generation from day to day, within the week). These seem reasonable assumptions and they are normally assumed because they simplify the whole problem of optimal operational planning.

In the case of no hydro storage, (4), (5), (6), (7) and (8) reduce to:

$$U_i = F_i + M_i \tag{9}$$

$$t_{min} \le \sum_i T_i \le t_{max} \qquad \text{(8) rep.}$$

To optimise it is necessary to know the cost function $C_i = f_i(U_i)$ and the revenue function $D_i = g_i(M_i)$, with C_i and D_i expressed in monetary units (L or \$).

In Sec. 7 some considerations will be developed about these functions.

The optimality condition in this case will be :

$$C = \sum_i (C_i - D_i) = \min \tag{10}$$

5.2 Second case : no thermal generation (or "purely hydro")

In this case (4), (5), (6), (7) and (8) reduce to:

$$I_i + H_i = F_i + M_i \quad \text{for any i} \tag{11}$$

$$V_i = V_{i-1} + A_i - H_i \quad \text{for any i} \qquad \text{(5) rep.}$$

$$0 \le V_i \le V_{max} \quad \text{for any i} \qquad \text{(6) rep.}$$

$$a_{min} \le \sum_i H_i \le a_{max} \qquad \text{(7) rep.}$$

The optimisation requires a knowledge of the functions $C_i' = f_i(I_i)$ and $D_i = g_i(M_i)$.

The condition for optimum is

$$C = \sum_i (C_i' - D_i) = \min \tag{12}$$

5.3 Third case : hydro-thermal power system

This is the most general case, and the problem has already been formulated by (4), (5), (6), (7) and (8). The condition for the optimum is again

$$C = \sum_i (C_i - D_i) = \min \qquad \text{(10) rep.}$$

5.4 Fourth case : no market opportunity

In this case there is no secondary load, no opportunity import or export, and hence M = 0, U = T.

The problem may now be formulated as follows:

$$H_i + T_i = F_i \quad \text{for any i} \tag{13}$$

$$V_i = V_{i-1} + A_i - H_i \quad \text{for any i} \qquad \text{(5) rep.}$$

$$0 \le V_i \le V_{max} \quad \text{for any i} \qquad \text{(6) rep.}$$

$$a_{min} \le \sum_i H_i \le a_{max} \qquad \text{(7) rep.}$$

$$t_{min} \le \sum_i T_i \le t_{max} \qquad (8) \text{ rep.}$$

$$C = \sum_i C_i'' = \min \qquad (14)$$

with $C_i'' = f_i(T_i)$.

Many utilities may recognise this case as their own problem; as a matter of fact, in many cases the secondary load concept is not used, and opportunity exchanges are considered as parameters and not as true variables.

The reason for this last assumption may be that, on the one hand, the introduction of a new variable makes it more difficult to solve the problem; and on the other hand, often the purchase or sale opportunities are at prices well below (or above) their own production costs, so that it is not necessary to make sophisticated evaluations to assess the economy of trading.

However, it may happen that the prices are near the production costs; in this case a check may be made by running the optimisation problem (13), (5), (6), (7), (8) and (14) twice, that is, with and without the new exchange contract, which is considered as a "must take" or a "must sell", and comparing the two operation costs (this is the meaning of "considering the opportunity exchanges as parameters and not as true variables").

To have the exchange opportunities as variables would perhaps require, in addition to a more complicated optimisation procedure, stricter co-ordination between the interested utilities.

5.5 Fifth case: purely hydro, no opportunity exchanges, no secondary load

A particular case deserving attention (included in the second and fourth cases above) is that in which U = M = 0:

$$H_i = F_i \quad \text{for any i} \qquad (15)$$

$$V_i = V_{i-1} + A_i - H_i \quad \text{for any i} \qquad (5) \text{ rep.}$$

$$0 \le V_i \le V_{max} \quad \text{for any i} \qquad (6) \text{ rep.}$$

$$a_{min} \le \sum_i H_i \le a_{max} \qquad (7) \text{ rep.}$$

This could be the case of a non-interconnected, purely hydro utility, feeding firm load only. Clearly, in this case the degree of freedom is zero and hydro generation must rigidly follow the firm load; depending on the series of values A_i, and on the value V_{max}, there may be in some MTIs unavoidable spillages, or firm load curtailment.

Some kind of optimisation could however be carried out at the lower level (short term), taking into account the efficiency curve shapes of the various power plants in order to minimise the water consumption for any global energy output.

6. Value of water

It has been assumed that hydro energy has zero operation cost (note: depreciation costs must not be considered in operation problems); on the contrary, a cost has been associated to thermal generation and to energy market availabilities, through the functions C_i and D_i.

6.1 Medium term water value

With the exception of the fifth case above (sec. 5.5), in which there is no optimisation problem, in the other cases hydro energy H may be given a fictitious operation value in the following manner (consider, for the sake of simplicity, the fourth case; the following considerations may be generalised to the more general third case):

* suppose that H_1, H_2, ..., H_k,....., H_n and T_1, T_2, ..., T_k,...., T_n are the optimum values, that is, those satisfying (14) and complying with the constraints (13), (5), (6), (7) and (8) (F_i and A_i are supposed to be known forecasted values);
* consider an elementary energy step e (that is, a value sufficiently low with respect to any H, T, F and A);
* suppose that in the time interval k the hydro generation departs from the optimal value H_k to $H_k + e$; correspondingly, T_k will move to $T_k - e$;
* repeat the optimisation (equations (13), (5), (6), (7), (8) and (14)), but keeping constant hydro and thermal generations of the k-th MTI at the values $H_k + e$ and $T_k - e$ respectively. The total cost will now be $C_{min} + d_k$, with $d_k > 0$;
* then the incremental water value in the k-th MTI is

$$w_k = \frac{d_k}{e} \left(\frac{\$}{\text{MWh}}\right) \tag{16}$$

The incremental value defined above is a substitution value, because the increment d_k is due to the variation of the generation schedules of thermal units (and to variation in the schedules of import/export in the more general case); hence it depends on the cost of the other sources (fuel, opportunity contracts). It depends not only on the costs prevailing in the particular MTI to which it refers, but also on the costs of all other MTIs constituting the period of study.

The value w defined above may be considered as a deterministic "medium term incremental water value". To account for randomness requires more sophisticated mathematical approach, the basic concepts remaining nevertheless unchanged (the "expected water value" instead of the "water value" will be the result; see also

Appendix D).

6.2 Short-term water value

Another definition which may be assumed is the "short-term incremental water value" v_j (\$/MWh).

Its definition may be given for two different contexts: hydrothermal (v_j') and purely hydro (v_j'').

For the first case (hydrothermal) suppose that H_1 (recall sec. A.3. for the definition of H) is the optimum hydro generation in the first MTI, as determined by (4), (5), (6), (7), (8) and (10); the first MTI is the next week. The short term scheduling is done with H_1 as constraint; that is, the total hydro generation in the next week will be equal to H_1 (the hypothesis assumed that A_1, water inflow, and F_1, firm energy, are forecasted exactly). This hydro generation will be distributed along the m ETIs of the week in the most economic manner, by the short term scheduling procedure.

Let λ_j (\$/MWh) be the thermal (plus opportunity import if any) incremental cost in the j-th ETI, as it results from short term optimisation.

Suppose also that in the same ETI a hydro power station is running at a flow of Q_o (cubic meters per hour), with a corresponding output P_o (MW).

The criterion followed to define the value of the water is a simple substitution criterion: if an increment ΔP in hydro generation is performed, an equal decrease ΔP will be obtained in thermal generation (or opportunity import), with a saving equal to $\lambda_j * \Delta P$ (in \$/h). Hence the incremental water value will be $\lambda_j * \Delta P$; on the other hand, since to produce $P_o + \Delta P$ an increment ΔQ is necessary to the flow Q_o, the incremental water value may also be expressed as $\mu_j * \Delta Q$.

Then $\mu_j * \Delta Q = \lambda_j * \Delta P$ and $\mu_j = \lambda_j * \Delta P / \Delta Q$.

This is expressed in \$ per cubic metre.

Once it is agreed to express in \$/MWh it is sufficient to divide μ_j by a constant reference value (for example, the average specific energy content c_{av}, which is expressed in MWh per cubic metre):

$$v_j' = \lambda_j \cdot \frac{\Delta P}{\Delta Q} \cdot \frac{1}{c_{av}} \tag{17}$$

Fig. A1 shows the trends of P versus Q and of $\Delta P/\Delta Q$ versus P for a typical hydro plant.

For the second case (purely hydro), the following simple definition may be assumed:

$$v_j' = w_1 \cdot \frac{\eta}{\eta_{max}} \tag{18}$$

In eq. (18) w_1 is the medium-term incremental water value considered above, prevailing for the first MTI; η is the efficiency of the plant as a function of output, and η_{max} is the maximum value of η (see Fig. A2).

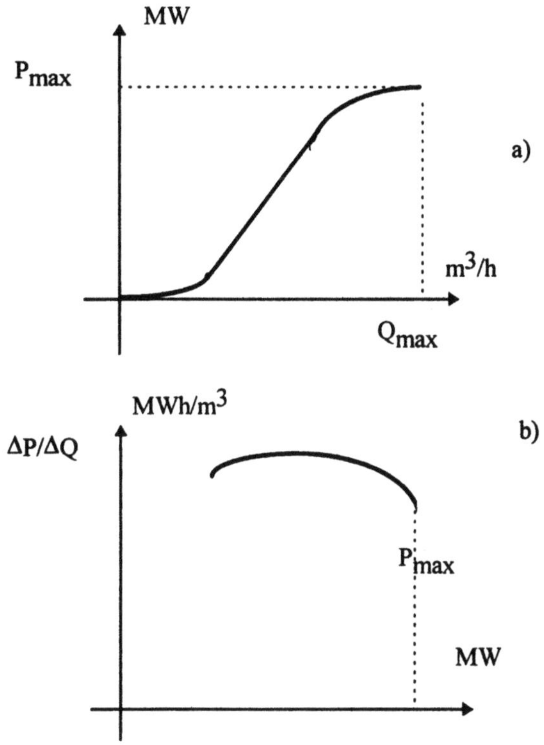

Figure A.1: *Trend of output P (MW) versus input flow Q (cubic metres per hour) (a), and of ΔP/ΔQ versus P (b), for a typical hydro unit.*

The introduction of the coefficient η/η_{max} which is less than or equal to 1, is intended to discourage non-optimum running of the hydro plants.

Both values, v_j' and v_j", may be useful when contracting unforecasted market opportunities, to evaluate their convenience, provided that they consist of small amounts of energy with respect to those given by the medium-term scheduling; in particular, v_j' and v_j" would perhaps become meaningless if the unforecasted market opportunities entail the shut-down or start-up of units.

Their use, in the above specified conditions, avoids having recourse to time-consuming re-computation of the medium-term optimal schedules, to get (theoretically) more exact economic evaluation of the unforecasted market opportunities.

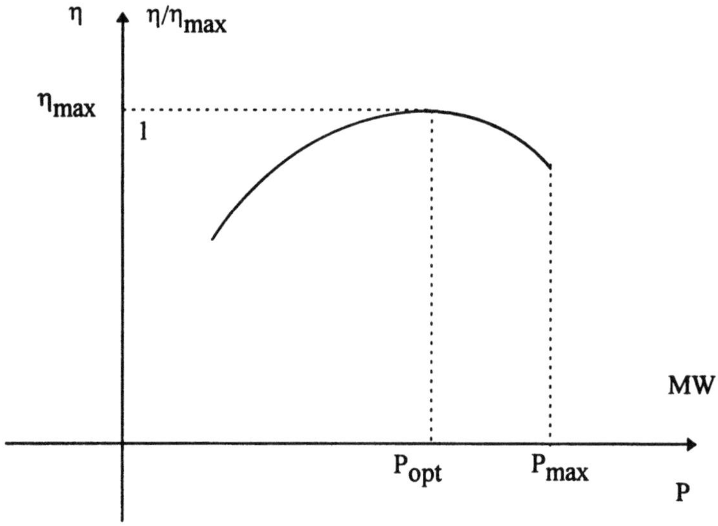

Figure A.2: *Efficiency versus output for a typical hydro unit*

7. Cost function and revenue function

7.1 Thermal generation plus opportunity import (cost)

To take into account all aspects required to give a complete definition of the cost function for thermal generation and opportunity imports (including start-up and shut-down, and cycling of thermal units), the duration of the MTI to be considered is at least one week. This is due to:

- long start-up and shut-down times (many hours in the case of most thermal units);
- start-up and shut-down costs (for example, the penalty factor associated with starting a cold generator may be prohibitive, particularly if it is only scheduled to run for a very short duration);
- the adverse effects of cycling thermal plants.

In other words, with the present state of flexibility of thermal units, an MTI of one week (or more) allows complete evaluation of the operation cost with sufficient accuracy, with no need to consider external conditions (that is, conditions relating to adjacent MTIs).

Figure A.3 refers to a typical MTI and shows, arranged in increasing order, the specific costs of thermal generation (Fig. A.3a) and of imported energy (Fig. A.3b).

With regard to Fig. A.3a the increase in specific costs is due to the fact that as the requested thermal generation increases, less and less economic units shall be started-up. Fig. A.3a is supposed to take into account, perhaps in a simplified manner, the start-up costs and the fact that thermal units undergo, during the week, cycling according to their technical characteristics and following the shape of the load curve (for example, they will follow a flat generation profile as far as possible, based on past experience).

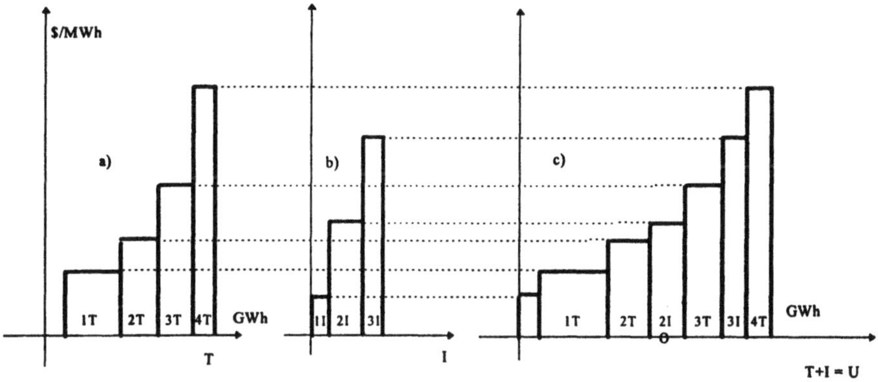

Figure A.3: *Specific costs, arranged in increasing order, of various blocks of: thermal generation (a), opportunity purchase or import (b), and addition of the two.*

Fig. A.3b shows in a similar manner the specific costs of imported energy.

In Fig. A.3c the two sources are merged together, again in increasing order of specific cost.

It may be recognised that recourse will be taken in the example shown , first of all to the first block, or parcel of imported energy (1I); next to the first block of thermal generation (1T); then to the second block of thermal generation (2T); etc.

Of course, arranging the energy in an increasing order of specific cost is an obvious economic criterion; nevertheless, should some other reason call for a different order, this may well be adopted (but with drawbacks in economy).

Note that load curtailment may be represented by one or more equivalent thermal units, with appropriately high cost (in Fig. A.3 the block 4T could well represent a first block of load curtailment).

The determination of the specific costs of various blocks (especially those of thermal generation) may appear to be poorly approximated; on the other hand, the search for a higher accuracy (for example, to take better account of the shape of the thermal generation diagram during the week, or the penalties corresponding to start-up costs) would be meaningless, since the various parameters of the MTI under consideration are subject to uncertainty (higher and higher as the MTI moves into the future).

From Fig. A.3c the function $C_i = f_i(U_i)$ is easily built up (Fig. A4), this simply being the integral of the curve of Fig. A.3c.

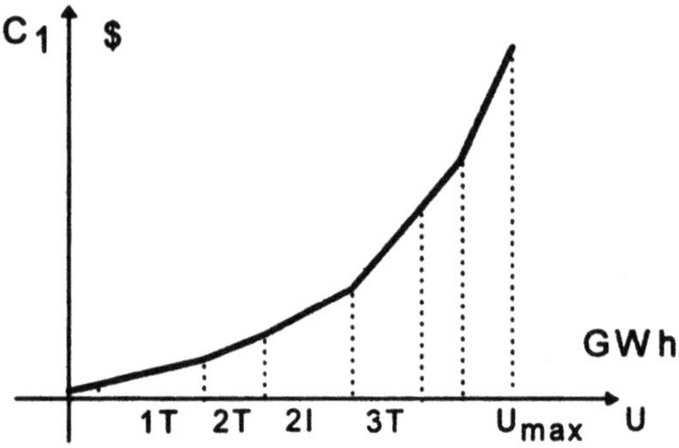

Figure A.4: *Cost function of the combination of thermal generation and opportunity purchase*

7.2 Secondary load plus opportunity export (revenue)

Fig. A5 shows, in a similar manner, the specific sale prices of secondary load L (A5a) and of opportunity export E (A5b), arranged in decreasing order; Fig. A.5c is the combination of the two.

The revenue curve $D_i = g_i(M_i)$ is obtained in the same way as C_i was obtained (Fig.A.6).

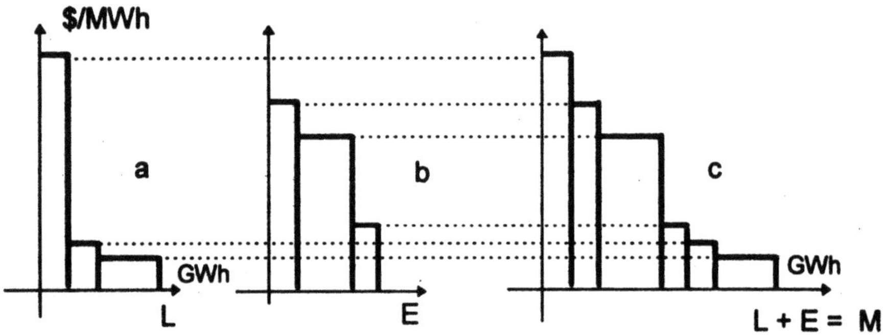

Figure A.5: *Specific sale prices, arranged in decreasing order, of various blocks of secondary load (a), opportunity export (b), and the merge of the two*

7.3 Combining the two functions

The two curves (Fig. A4 and Fig. A6) are defined separately because they represent two different parameters: C_i is a cost, D_i a revenue.

In the optimisation procedure care must be taken when dealing with cases in which both are present.

In the Appendix B some comments are made about this point.

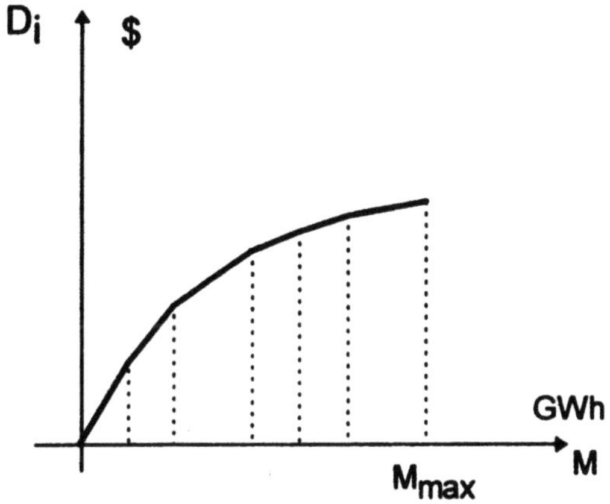

Figure A.6: *Revenue function of the merge of secondary load and opportunity export*

APPENDIX B

The problem is formulated in the case of sec. A.5.1 ("purely thermal") as:

$$U_i = F_i + M_i \quad \text{for any i} \qquad (9) \text{ rep.}$$

$$\sum_i \left(f_i\left(U_i\right) - g_i\left(M_i\right)\right) = \min \qquad (10) \text{ rep.}$$

In equation (8) is disregarded, as well as the other constraints appearing in the most general case A.5.3, in order to make more straightforward the illustration of the criterion of optimisation.

Of course, in practical cases some of those constraints are active, and their presence introduces some further mathematical concerns.

The solution of (9) and (10) is given by

$$\frac{df_i}{dU_i} - \lambda_i = 0 \qquad i = 1,..., n$$

$$-\frac{dG_i}{dM_i} + \lambda_i = 0 \qquad i = 1,..., n$$

$$U_i - M_i - F_i = 0 \qquad i = 1,..., n$$

In this case there are n separate problems (one for each MTI). The preceding equations may be rewritten

$$\frac{df_i}{dU_i} = \frac{dg_i}{dM_i} = \lambda_i \qquad i = 1, ..., n$$

$$U_i = F_i + M_i$$

They state the well-known "equal incremental costs" criterion, which in this case is applied to only two sources; the solution is graphically illustrated in Fig.B1 (the horizontal line is shifted up and down until an intercept equal to F_i is found). The two curves df_i / dU_i and dg_i / dM_i, are those of Fig. A.3c and Fig. A.5c, respectively.

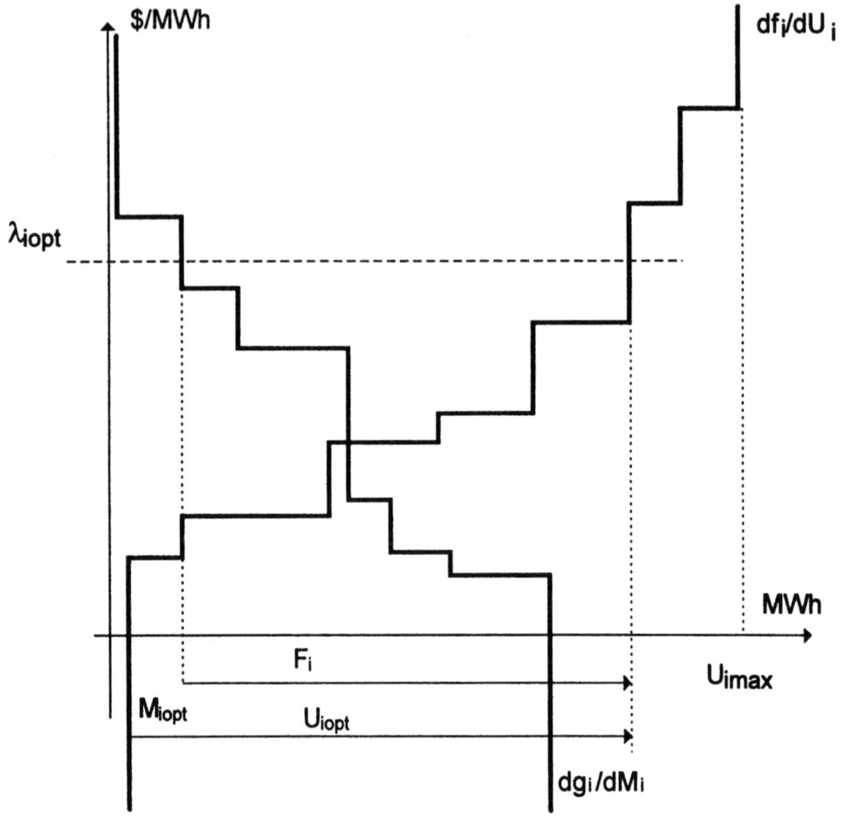

Figure B.1: *Finding the optimum values of M and U, with given F, in a typical MTI (M, U and F are defined in sec A.4)*

For a solution to exist,

$$U_{i\,max} - M_{i\,min} \geq F_i$$

APPENDIX C

In each ETI (elementary time interval of the short-term scheduling period considered) the optimal choice of P (thermal generation plus opportunity import) and S (opportunity export plus secondary load) must be performed, satisfying the load R.

P, S and R are expressed in MW.

The problem has the same approach as dealt with in Appendix B. It will be considered with the same limitations with which the last was considered.

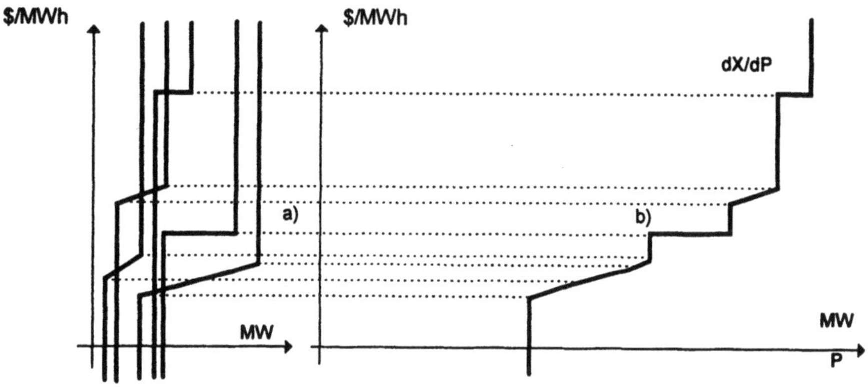

Figure C.1: *(a): incremental cost curves of thermal units and of blocks of imported energy; (b) : total incremental cost curve, obtained from (a) with the criterion of equal incremental costs. Reference to a typical ETI*

Fig. C1(a) shows the incremental cost curves of the running thermal units and of the available blocks of import; with the criterion of equal incremental costs, the curve of Fig. C1(b) is obtained, which will be called dX / dP ($/MWh), X being the total (minimum) cost ($/h), which is a function of P.

Fig. C2(a) shows the income curves ($/h) of the various available blocks of opportunity export and secondary load; Fig. C2(b) shows the corresponding derivatives (incremental revenues), and Fig. C2(c) the overall incremental revenue curve, obtained from that of Fig. C2(b) with the criterion of equal incremental revenues.

The curve of Fig. C2(c) will be called dY / dS ($/MWh), Y being the total (maximum) revenue ($/h), which is a function of S.

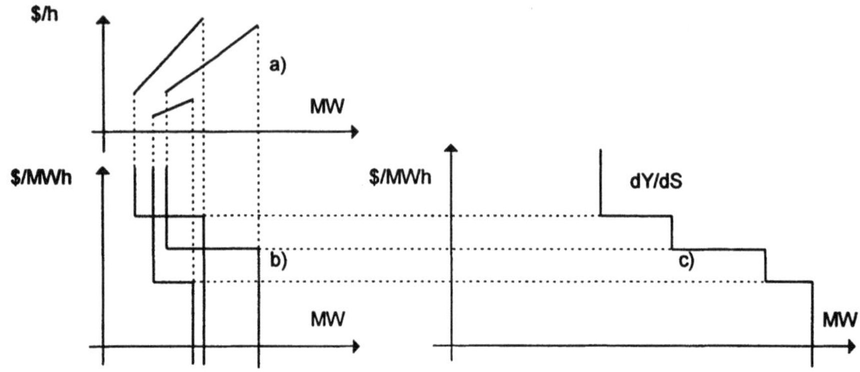

Figure C.2: *Income curves of various available opportunity exports and secondary loads, (b): incremental curves of those blocks; (c) : total incremental revenue curve, obtained from (b) with the criterion of equal incremental revenues. Reference to a typical ETI*

The problem is stated as follows:

$$\frac{dX}{dP} = \frac{dY}{dS} = \lambda$$

$$P = S + R$$

The solution is shown in Fig. C3.

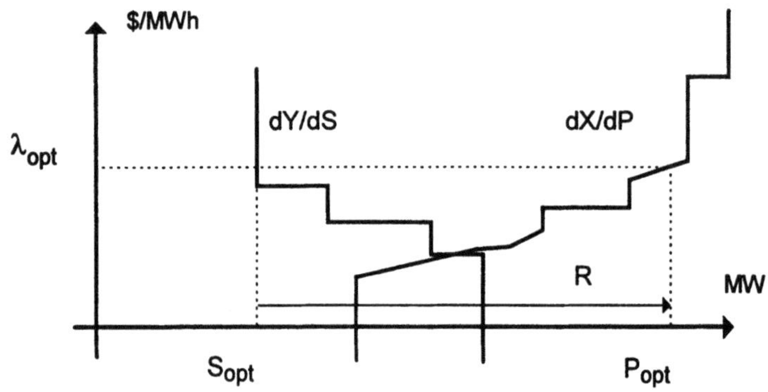

Figure C.3: *Finding the optimum values of S and P, with given R, in a typical ETI*

APPENDIX D

Water value computation (probabilistic case).

The concept of incremental water value relies on the following assumptions (recall that an assumption, or principle, is not proved, but simply accepted due to its likelihood):

I. water has no value in itself, hence a value can be assigned to it on the basis of equivalence or substitution; in particular, the incremental water value is assumed to be equal to the incremental cost of fuel or of any other resource which can replace it;

II. the incremental water value is a decreasing function of the amount of water stored.

In the example it is assumed that a period of study of one year consists of 12 elementary time intervals of one month.

The energy quantities are expressed as integers, in a common unit (for example 1 GWh or 1 TWh); the cost or value quantities are expressed as real numbers, in a common arbitrary unit ($/MWh or pounds/kWh or pesetas/GWh).

The power system is supposed to consist in a reservoir of maximum storage equal to 20; an equivalent hydro power station (fed by that reservoir) with maximum energy output varying from month to month; an equivalent thermal power station with maximum energy output varying from month to month and with incremental cost functions also different, in general, from month to month.

The main characteristics of the example power system are summarised in Table 1 and Fig. D1.

The only parameter considered as stochastic is water inflow: for each month a discrete number of water inflows have been considered, each one with its own probability (for example, in month 3: inflow 4 with probability 10%; 6 with 60%; 8 with 30%); the expected values shown in table 1 are the averages (for month 3: $0.1*4+0.6*6+0.3*8 = 6.4$).

By means of the methodology considered in sec. 6.3.3.2 (utility H), the "equi-value" (or "equi-price") curves of water storage shown in Fig. D2 have been determined. The figures near the curves shown represent the incremental water values, expressed in the same units as the incremental thermal costs of Fig. D1.

The shapes and parameters of the equi-price curves are strongly affected by the upper limits of hydro and thermal generation (max hydro and max thermal output of Table I) and by the distribution of the inflows along the period under study (last column of Table I).

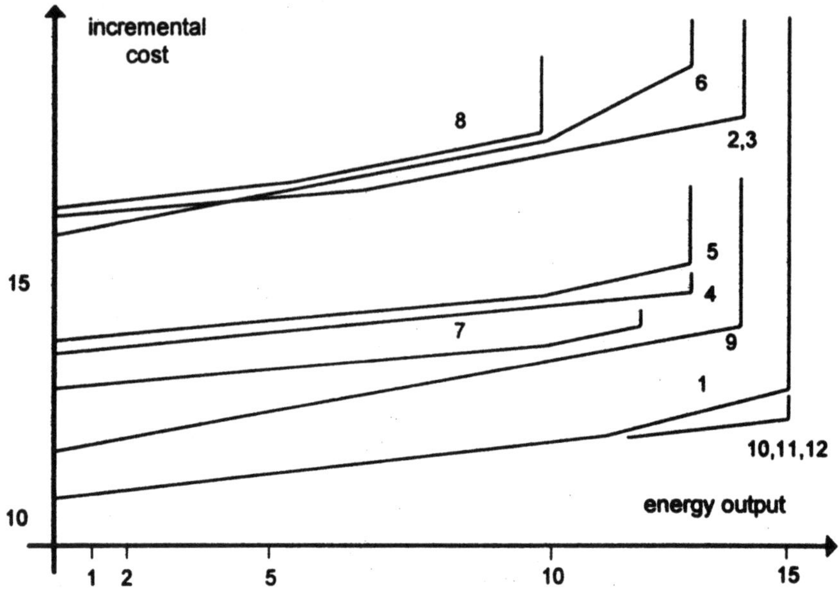

Figure D.1: *Incremental costs of thermal generation*

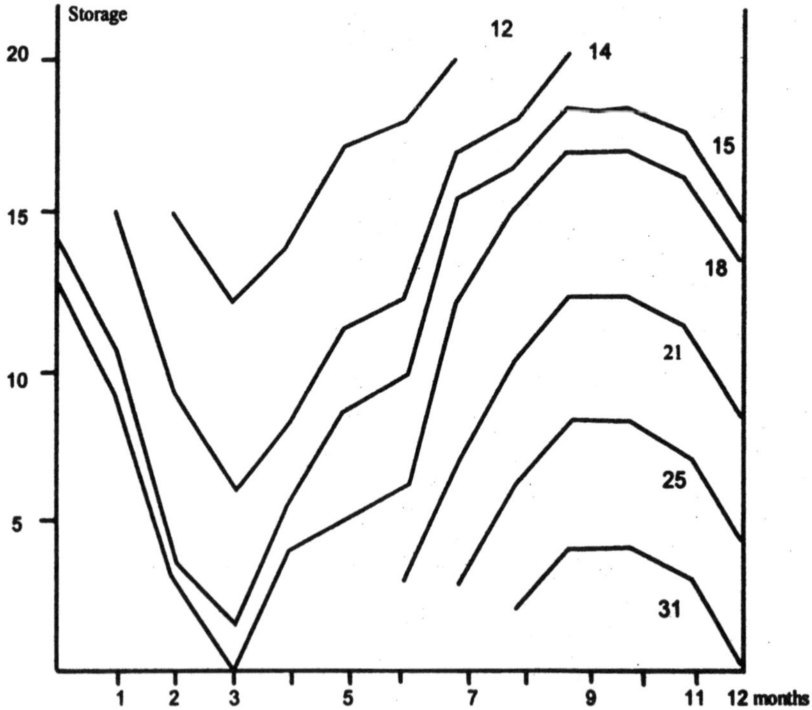

Figure D.2: *Equi-price curves of water, obtained for the example system*

Table I: *Generation, load and inflow parameters of the example power system*

Time stage	Max hydro output	Max thermal output	Load	Expected inflow
1	10	15	20	1.3
2	9	14	21	3.2
3	9	14	20	6.4
4	8	13	17	10.4
5	9	13	18	12.4
6	9	13	18	10.2
7	8	12	16	10.2
8	5	10	13	6.2
9	8	14	16	4.2
10	9	15	18	3.2
11	10	15	18	2.2
12	10	15	20	2.2

APPENDIX E

Terminology

Filling Period : the time required for filling the catchment structure from the lowest to the highest level normally allowable in use, with constant supply flow equal to the characteristic mean corrected flow; the downstream hydro power station being supposed to be at rest (UNIPEDE 1.2.09).

Run-of-the-river installation : catchment structure with filling period not higher than 2 hours (UNIPEDE 2.7.01).

Pond (or poundage) installation : catchment structure with filling period between 2 and 400 hours (UNIPEDE 2.7.02).

Reservoir installation : catchment structure with filling period above 400 hours (UNIPEDE 2.7.03).

Must run unit (hydro or thermal) : generating unit which shall run at a predetermined and constant output (positive).

Shall run unit (hydro or thermal) : generating unit which shall run at an output variable between a positive minimum (less than maximum output) and its maximum allowable output.

Scheduling period : the future period of time covered by forecast and scheduling activities of a particular procedure.

MTI (medium term elementary time interval, or medium term time stage) : the length of time interval in which the scheduling period is divided, when dealing with medium-long term operation planning. For example: a scheduling period of 3 years may be divided into 36 time intervals (MTIs) of 1 month duration each. Within the MTI all the data, parameters and unknowns are considered constant with respect to time, as far as the particular scheduling procedure is concerned.

ETI (short term elementary time interval, or short term time stage) : the same as MTI, but when the scheduling procedure is short term.

Integral type constraint (with respect to time) : a constraint which involves unknowns belonging to at least two ETIs (or MTIs).

Secondary load (or non-firm load, or opportunity load) : a load which is contracted with the clause of possibility of interruption (with advice), also in normal network conditions. A typical example of secondary load is electric heating, substituting fuel if the contracted electric tariff is economically advantageous (customers are supposed to be equipped with both facilities so as to be able to switch from one to the other in short time).

Secondary (or non-firm, or opportunity) import or export : an import or export which is contracted with the clause of possibility of interruption (with advice).

Cycle of pumped storage : the period of time during which the balance of water must be closed. For example, daily cycle is that in which the water pumped during night is turbined in the following day hours; a weekly cycle is that in which the water pumped during week-end is turbined in the following 5 working days. The possibility of performing a longer cycle depends on the relative magnitude of the upper and lower reservoirs or ponds, with respect to installed pumping and generating capacity.

INDEX

—I—

—L—

—M—

—N—

—O—

—P—

—Q—

—R—

—S—

—T—

—U—

—W—

MIX
Papier aus verantwortungsvollen Quellen
Paper from responsible sources
FSC® C105338

If you have any concerns about our products,
you can contact us on
ProductSafety@springernature.com

In case Publisher is established outside the EU,
the EU authorized representative is:
Springer Nature Customer Service Center GmbH
Europaplatz 3, 69115 Heidelberg, Germany

Printed by Libri Plureos GmbH
in Hamburg, Germany